Apparel Finishing
and
Clothing Care

Apparel Finishing
and
Clothing Care

Edited by

Dr. M. Parthiban

Dr. M. R. Srikrishnan

WOODHEAD PUBLISHING INDIA PVT LTD

New Delhi

Published by Woodhead Publishing India Pvt. Ltd.
Woodhead Publishing India Pvt. Ltd.,
303, Vardaan House, 7/28, Ansari Road,
Daryaganj, New Delhi - 110002, India
www.woodheadpublishingindia.com

First published 2020, Woodhead Publishing India Pvt. Ltd.
© Woodhead Publishing India Pvt. Ltd., 2020

Woodhead Publishing India Pvt. Ltd. ISBN: 978-93-88320-23-8
Woodhead Publishing India Pvt. Ltd. e-ISBN: 978-93-88320-24-5

Typeset by Bhumi Graphics, New Delhi
Printed and bound by Replika Press Pvt. Ltd.

Contents

Preface

Apparel Finishing is a process used in manufacturing of fiber, fabric, or clothing. In order to impart the required functional properties to the fiber or fabric, it is customary to subject the material to different type of physical and chemical treatments. For example wash and wear finish for a cotton fabric is necessary to make it crease free or wrinkle free. In a similar way, mercerising, singeing, flame retardant, water repellent, water proof, antistatic finish, peach finish, etc... are some of the important finishes applied to textile fabric.

Broadly it can be classified into following classes, which are used individually or in combination with each other. It could be categorized as wet finishing, dry finishing, durable finishes and non durable finishes. Apparel Finishing is a term commonly applied to different process that the textile material under go after pretreatment, dyeing or printing for final embellishment to enhance their attractiveness and sale appeal as well as for comfort and usefulness.

Similarly, dirty clothes can harbor microorganisms and wearing the same can lead to skin infections. Body odor can also occur wearing clothes with the bacteria and fungi found on them. There forth, keeping your clothes clean and fresh is a primary importance in our day to day lifestyle. It helps in knowing how to take care of a certain type of fabric in terms of washing, drying or ironing. It is important that if you run a textile or clothing business to always include the care label with your clothing. While not all retailers require a care/content label to be sewn in the garment. The washing process is famously tough on clothes. Agitating, tumbling, and coming in contact with other garments can leave garments faded, stretched, pilling, and damaged. Unless your outerwear is visibly dirty, you probably don't need to wash it after each wearing.

Hence, this book will provide the technological concepts related to different types of finishes, selection of chemicals for different types of fabrics and application of these finishes on to the fabrics and garments. This book also addresses the different denim finishes and various methods for imparting desired effect on denim garments. The significance of wash care labels have also been dealt in detail.

Dr. M. Parthiban
Dr. M. R. Srikrishnan

1

Wash care of apparel

Dr. Subrata Das

Professor (Fashion Technology), Bannari Amman Institute of Technology, Sathyamangalam, Erode District, Tamil Nadu- 638401. E-mail: subrata40in@yahoo.co.in

Abstract : Care instruction of fabric and apparel has been the subject of many recent discussions. It is not only important to increase the life of the garment but also it is a prime requirement for the export market. Without proper care, the apparel loses both aesthetic and technological values required in the fashion industry. In the present discussion, wash care fundamentals of apparel has been discussed. Wash care label verification process has been highlighted. Wash care of different apparels has been investigated through the case study.

Improper use of cleaning garment may lead to the deterioration of inherent properties. Thus, different possibilities of garment damage during cleaning have also been focused.

Keywords : Washing, drying, ironing, bleaching, dry cleaning, dimensional stability, colour fastness

1.1   Introduction

Functional and aesthetic needs of the people are addressed in clothing technology which, however, need to be retained for its useful lifetime. Purchasing of apparels and wearing decisions of consumers are traditionally been influenced by social, psychological, physiological, physical, cultural and economic parameters. Of late, awareness of the impact of anthropological activities on the environment has been seriously thought of and as a result environmental factor finds its place in the decision making process. Apparel and textiles appear dull and dingy during normal use. Due to economical reason, apparel and textiles must be cleaned and refurbished for reuse without modifying their functional and aesthetic properties. Consumers have the choice to clean apparel at home or at professional dry cleaning establishments. Regulatory bodies associated with the apparel industry throughout the world emphasized the importance of the fact that apparel items should have a permanent care label, which provides verbal information about regular care. The aim is to protect the interest of the consumer through accurate care information to extend the useful life of garments.

The study outlines the intricacies of wash care and addresses the difficulties encountered in defining reliable care instructions. Conceptual wash care spectra for non-aqueous and aqueous cleaning process are discussed and technology options, cleaning mechanisms, property issues, and garment damage potentials have been highlighted. Wash care label verification steps have been investigated and case studies of different export-oriented apparels for reputed brand buyers are discussed.

1.2   Ash care fundamentals

1.2.1   Technology options

The mechanism of cleaning of apparels entails the breaking of the soil-textile interaction forces to loosen and transport the heterogenous soils away from the substrate. The soils should be concentrated for proper disposal, preferably as non-hazardous waste. But, the process must clean clothes to satisfy consumer needs, and it must be economically and environmentally feasible. There exist two practical boundary technologies: non-aqueous and aqueous cleaning.

Non–aqueous cleaning: Non-aqueous textile cleaning technology is governed by the properties of textiles and soils. Though perchloroethylene is a proven medium for professional textile cleaning, any other non-polar media, such as petroleum, carbon dioxide or other non-polar liquids, which meet the textile cleaning performance requirements, could be used. Historically Stoddard solvent, Carbon Tetrachloride, and Valelene 113/ Freon 113 were used as dry cleaning solvents. Recent reagents include high flash point hydrocarbons DF-2000 (140°F flash point), modified hydrocarbons blends (Pure Dry), glycol ethers (Dipropylene glycol tertiary-butyl ether) (Rynex), cyclic silicone (Decamethylcyclopentasiloxane) [Green Earth] and supercritical carbon dioxide. Perchloroethylene is undoubtedly the most commonly used solvent with unmatched cleaning performance though it is considered as a persistent and bio-accumulative chemical that is toxic to the environment. Detergents are utilized in drycleaning solvents for enhanced cleaning capability. Sometimes, sizing chemical is added to restore garment shape, body, and texture.

Aqueous textile cleaning:At the other end of the spectrum is aqueous cleaning wherein the advanced professional wet cleaning technology makes it feasible to clean many textiles, which are traditionally cleaned in non-aqueous media. Wet cleaning is a non-toxic, environmentally safe alternative to dry cleaning, utilizing computer-controlled washing machines, biodegradable soaps and conditioners, and various types of pressing equipment that may be specialized for many different fabric and fiber types. Generally, aqueous cleaning involves processes like washing, bleaching, drying, and ironing. Mild to harsh treatments are associated with each of the processes. Determination of the severity of the condition of treatment is dependent upon the quality of the substrate and the intended end use of the consumer. Wet cleaning offers several advantages, such as lower costs for start-up capital, supplies, equipment, and hazardous waste disposal, as well as less reliance on skilled labor. It does not use hazardous chemical, does not generate hazardous waste, nor create air pollution, and reduces the potential for water and soil contamination.

1.2.2 Cleaning mechanism

Non-aqueous media in colloid systems ensure satisfactory cleaning of apparels. In non-aqueous media, polar media is made up of molecules that have separate centers of positive and negative charge creating a dipole moment whereas non-polar media is made up of molecules that have a more symmetrical distribution of electron, such that no dipole moment is created. Polar soils are better removed in water than non-polar solvents and that non-polar soils are easier removed in non-aqueous solvents. Professional textile cleaners can optimize soil removal if they have access to both media.

1.2.3 Nature of substrate

The care procedures to which apparel is subjected would determine how well the fabric continues to retain its colour, appearance, hand, drape, durability, etc. The chemical structure and composition of the fibres comprising the yarns used for the fabrics engineered in apparel determine the procedures for refurbishing.

Chemical sensitivity of the fabric would influence the type of detergent, bleach, or drycleaning treatments to be used when washing or drycleaning and the heat sensitivity would determine the temperature to be used in washing, drying or pressing the fabric. The chemical composition of the fibres in the fabric must be correlated with the appropriate care treatment to maintain the garment or textile article in optimum condition.

In addition to chemical composition, the way in which the fibres are spun into yarns and the yarns are woven/knitted into fabrics will also affect care procedures. A fabric, which is loosely woven with yarns that are loosely twisted, will not be dimensionally stable. Special precautions need to be taken in construction, wear and care of the fabrics so that they maintain their shape and to extend the serviceability of the fabric goods. The construction of the garment or textile product adds to the complexity of the care procedures. For instance, garments that are colour blocked or have trims of different colours, even though the fabrics are washable may have to be drycleaned or washed and spin-dried immediately to ensure that there is no bleeding of the colours to produce self-staining/cross-staining.

Finishes, colour, and design of a garment are important parameters in wash care. Various easy care finishes are now available in the market. It is difficult for the consumer or producer to understand how a fabric should be maintained by simply looking at it. Most finishes are frequently non-visible and fibre content in composite fibre fabrics are usually not all that easy to determine. The consumer has to depend on information given on the fabric/garment label. The producer of garments and textile articles need to indicate accurate fibre content and care instructions on labels so that the consumer may be correctly guided in the care of the items.

1.3   Washcare label verification

1.3.1   Purpose

To verify the care label submitted by the client with the reference of the standards such as ISO 3758: 1991 (1), ASTM D3136:2000 (2), ASTM D5489-96C (3), FTC Care Labelling guide (4) and 16 CFR part 423 (5).

1.3.2   Theory

Care labeling regulation 16CFR423 accepts care-labelling symbols designated in ASTM D5489-96c.

Following should be the order of reporting care symbols

ASTM/CFR	ISO
Washing	Washing
Bleaching	Bleaching
Drying	Ironing
Ironing	Drycleaning
Drycleaning	Drying

In case of washing, drying and ironing symbol dots refer to the following temperatures:

Washing

Dots	ASTM		ISO
.	30 °c	Cold	30 °c
..	40 °c	Warm	40 °c
...	50 °c	Hot	50 °c
....	60 °c	Very Hot	60 °c
.....	70 °c	Very Hot	70 °c
......	95 °c	Very Hot	95 °c

The numerical (temperature in °c) or the dot system or both can be used to represent the maximum water temperature. As per 16 CFR 423, when no temperature is given, hot, warm, or cold term can be used.

Hot - Water temperature ranging from 45°C to 63 °C.

Warm - Water temperature ranging from 31°C to 44 °C.

Cold - Temperature 30°C.

In addition to the basic symbols, a bar under the symbol means that the treatment is more gentle which is described as follows:

Symbol	ISO	AATCC
	Normal cycle	Normal cycle
	Gentle Cycle	Permanent Press
	Very gentle or wash as wool	Delicate/Gentle cycle

Drying Tumble dry:

Dots	ISO	ASTM/CFR
.	Low	Low
..	Normal	Medium
...	-	High

For US other drying instructions should read as:

| Line dry /
hang to dry | Drip dry | Drip.flat | Do not dry
(used with do
not wash) |

Ironing

Dots	ISO	ASTM/CFR
.	110 °c	110 °c/Cool
..	150 °c	150 °c/warm
...	200 °c	200 °c/Hot

1.3.3   Procedure

1.3.3.1   *Test to be performed on applicant's provided care instruction for USA market:*

- If any symbol is not given in care, instruction indicates that the submitted sample is safe for that instruction. Therefore, that instruction should be verified at the severest condition.
- If washing instructions are given, verify the tests that relate to washing. No need to check for dry cleaning instructions.
- If the instructions say, do not dry clean, then there is need to test the dry cleaning tests.

(I) Given - Machine Wash /Hand wash label without warning: (4 symbols - wash, bleach, drying and iron)

1. Dimensional stability to washing:
   - Test according to the applicant's provided care label/washing conditions. (e.g. washing temperature, washing cycle, drying method).
2. Colour fastness to washing (at fabric stage):
   - Test according to the applicant's provided care label / washing conditions. (e.g. mainly on washing temperature).
3. Colour fastness to chlorine bleach (in house method):
   - Exception on wool/animal fibres, silk spandex/elastic, and their blends.

4. Colour fastness to non-chlorine bleach (in house method).
5. Appearance after wash:
   • Test according to applicant's provided care label and assess the appearance after washing, drying and ironing.

Report staining on multifibre after first wash in case of product testing.

Comment on spirality is mandatory (fabric & garment) and to be reported in percentage.

Temperature conditions		AATCC 61 program		Dimensional change conditions	
Words	°c	Hand wash	m/c wash	3 cycles	
Cold	30	1 A (30)	2A (30)	27 ± 3 °c	Temperature is specified then performs testing at that particular temperature.
Warm	31-44	1A (40)	2A (49)	40 °c	
Hot	45-63		3A (71)	60 °c	

Give the below comment if care instructions refer to 30°C

Note: "Test performed as per applicant provided care instructions"

(II) Machine wash/Hand wash with a warning:

Test items to be conducted are mainly according to (I) Machine wash/Hand wash label without warning.

However, the washing condition during testing may be varied or added when subjected to the following warnings:

1. Do not bleach / No bleach:
   • For wool/animal fibres, silk, spandex/elastic, and their blends, give comment "Bleach is not required due to inherent properties of silk/wool/... fibres".
   • For other items, test for both colour fastness to chlorine and non-chlorine bleach.
2. Do not tumble dry:
   • For wool/animal fibres, silk, spandex/elastic and their blends, we will follow the drying instruction as stated (such as line dry, flat dry, etc) and not necessary to test for tumble dry low.
   • For other items, tumble dry should be conducted first except for hand wash care label.
3. Do not dryclean:
   • Check colour fastness to drycleaning including all embellishments and materials.

4. Do not Iron:
   - No need to check for ironing instructions for crinkled styles.

(III) "Dryclean only" label: (Symbol: Either m/c wash crossed and dryclean open or written as only Dryclean)

1. Dimensional stability to washing:
   - Test according to Hand wash Cold and line dry/ flat dry (depending on product type).
2. Dimensional stability to commercial drycleaning.
3. Colour fastness to washing(at fabric stage):
   - Test according to hand wash cold condition.
4. Colour fastness to drycleaning with perchloroethylene.
5. Colour fastness to chlorine bleach (IHTM method):
   - Exception on wool/animal fibres, silk, spandex/elastic, and their blends.
   - Comment to be given "Bleach is not required due to inherent properties of silk/wool/… fibres."
6. Colour fastness to non-chlorine bleach (IHTM method).
7. Appearance after washing:
   - Assess appearance after hand wash cold, line/flat dry, ironing.
   - Attach multifibre in case of product testing & report the staining on multifibre also.
8. Appearance after commercial drycleaning.

1.3.3.2   *Test to be performed for Comment on Applicant's provided Care*

Instruction for U.K. and European Market–

(I) Care Label without Warning:

Exception for "Do not bleach or warning label

1. Dimensional stability to washing: (Follow ISO 6330:1984)

Test according to the applicant's provided care label/washing conditions. (e.g. washing temperature, washing cycle, drying method)

(I) Colour fastness to washing:

1. Test according to the applicant's provided care label/washing conditions (e.g. mainly on washing temperature, "Machine wash" program or "Hand wash" program)

(II) Colour fastness to Bleaching with Hypochlorite (only on whites):

- Exception on wool / animal fibres, silk spandex/elastic and their blends.

  (i) Colour fastness to Drycleaning

  (ii) Appearance after commercial drycleaning – Only for delicate/ embellished styles.

  (iii) Care Label with Warning:

  Test items to be conducted are mainly according to (I) Care label without Warning.

  However, the washing condition during testing may be varied or added when subjected to the following warnings:

  1. Do not tumble dry:
     - For wool/animal fibres, silk, spandex/elastic, and their blends, we will follow the drying instructions as stated or line dry for woven or screen dry for knitting, and not necessary to test for tumble dry.
     - For other items, washing shrinkage after tumble dry should be conducted first except for hand wash care label provided, in order to check if the product is lower labeling.

  2. Do not dryclean:
     - Colourfastness to drycleaning to be conducted including all accessories.

  3. Do not Iron:
     No need to verify ironing instructions in crinkled styles.

  (iv) "Dryclean only" label: (Rest other symbols crossed)

  Colour fastness to Washing:
  - Test according to hand wash condition (30 °C/40°C)

  1. Dimensional stability to Washing:
     - Test according to simulated hand wash program conditions (30 °C/40°C)

  2. Appearance after wash and ironing.

  3. Colour fastness to Chlorine bleach with hypochlorite- only for whites
     - Exception on wool/animal fibres, silk spandex/elastic and their blends.

  4. Colour fastness to drycleaning

5. Dimensional stability to commercial drycleaning
6. Appearance after commercial drycleaning

Other Care Instructions:
- Spot clean- Applicable only for products which can not be washed or drycleaned such as PVC coated mats, wooden placemats, mats with Bamboo sticks, etc.
- For verification, only the appearance test is to be performed. Colour change or any other surface appearance change is to be assessed.
- Vacuum clean – Applicable only for products which can not be washed or drycleaned. It is generally given for floor coverings.
- For verification assess excess fibre removal or any change observed.

Note:
- For Crinkled styles: Drying tag (twist, wring or crush) should be asked from the supplier.
- For Crush, drying bag should be supplied by the supplier.

1.3.4   Reporting Instruction:

Results and requirements are to be reported for all tested parameters.
- If sample verifies to applicant provided care instruction then report the same care label on the front sheet.
- If the sample passes to warning instructions (do not bleach, do not dryclean) for US market then report actual observation as "Verified/ Recommended care instruction: m/c wash…, only non-chlorine bleach if needed, …..dryclean
- If the sample passes to warning (do not bleach) for European market then report\ it as "do not bleach" only.

1.4   A case study on wash care

A case study was conducted in the ready-made garment industry in Bangladesh for the reputed retailers in the globe based on the merchandise submitted to Bureau Veritas Consumer Product Services Lab in Dhaka, Bangladesh (a third party multinational organization engaged in fabric/apparel testing services) prior to export by different factories associated to bulk shipment. Table 1 depicts that, based on the fibre content and the type of the substrate how the care label instruction varies.

Table 1.1: Care instruction of different merchandises

Sl. No.	Fibre Content	Garment/Fabric	Name of the Company	Evaluated care instruction
1	70% Acrylic, 30% Wool	Ladies Long Sleeve Cardigan	SAKS INC. USA	Hand wash cold, Wash separately or with like colours, only non-chlorine bleach when needed, reshape and dry flat
2	100% Cotton	Toddler Twill Shortall	SAKS INC. USA	Machine wash warm, wash separately or with like colours, only non-chlorine bleach when needed, tumble dry low, warm iron when needed.
3.	100% Cotton	Toddler Denim Shortall	SAKS INC. USA	Machine wash cold, wash separately or with like colours, only non-chlorine bleach when needed, tumble dry low, warm iron when needed.
4.	100% Cotton	Canvas Cargo Shorts	SAKS INC. USA	Machine wash cold, with similar colours, only non-chlorine bleach when needed, tumble dry low, remove promptly, warm iron if needed.
5.	65% Polyester, 35% Cotton	L/S Aviator Shirt (Poplin)	Brylane, USA	Machine wash warm with like colours, use only non-chlorine bleach, tumble dry medium, medium steam iron.
6.	60% Cotton, 40% Polyester	Poplin CVC Big Shirt	Brylane, USA	Wash dark colours separately, machine wash warm, use only non-chlorine bleach, tumble dry medium, medium steam iron.
7.	100% Cotton	Sheeting L/SLV Pigment Dyed Shirt	Brylane, USA	Wash dark colours separately, machine wash warm, use only non-chlorine bleach, tumble dry medium, medium steam iron.
8.	60% Cotton, 40% Polyester	Poplin CVC Big Shirt	Brylane, USA	Machine wash warm, with like colours, do not bleach, tumble dry medium, remove promptly, medium steam iron.
9.	100% Cotton	Y/Dyed Stripe Long Sleeve Men's Shirt	VF Asia, HongKong	Machine wash cold, with like colours, only non-chlorine bleach when needed, remove immediately, tumble dry low, medium iron.

Contd...

Contd...

Sl. No.	Fibre Content	Garment/Fabric	Name of the Company	Evaluated care instruction
10.	88% Cotton, 12% Polyester	Men's L/S Shirt	PVH, USA	Machine wash warm, with like colours, only non-chlorine bleach when needed, tumble dry low, remove promptly, touch of with steam iron if desired.
11.	100% Cotton	Canvas Ladies Shorts/Skirts	Promod, France	Machine cold, gentle cycle, do not bleach, cool iron if needed, dryclean, do not tumble dry.
12.	100% Cotton	Fleece for Girls Tricot	Lindex, Hong Kong	Machine wash hot, normal cycle, do not bleach, hot iron, dryclean, tumble dry iron.
13.	100% Cotton	Boy's L/S T-Shirt with Emb. Applique	Lindex, Hong Kong	Machine wash hot, normal cycle, do not bleach, hot iron, dryclean, line dry.

1.5   Damage during cleaning of garment

Tailored or structured garments and light fashion items have often linings, trims, and other accessories or have complex designs features. They behave often differently in the same cleaning media. Damages to these items are less likely to occur in the non-aqueous medium than in the aqueous cleaning media. Thus, these garments are best cleaned in a non-aqueous medium. Many garments, such as overcoats, trousers, raincoats, or sweaters may be cleaned in either media. Shirts, blankets, sleeping bags, and linens are best wet cleaned. Occasionally, excessive polar or non- polar soiling dictates and overrides textile cleaning media selection criteria.

A deviation from care label instructions increases the risk of deteriorating garment quality with regard to dimensional stability, appearance, colour change, colour staining, and durability. The potential of a generation of such inferior characteristics in garments during cleaning is generally higher in the aqueous medium than in the non-aqueous medium. This is the major reason why today drycleaning is highly preferred for higher-end products in the apparel chain. Often, manufacturers mark their garments as "dryclean only" for an additional protection in refurbishing in order to restore desired quality acceptable to the consumer. Out of the different quality characteristics to be safeguarded in washing, dimensional stability is always regarded as an important parameter to the consumer world. The mechanism of dimensional stability is regulated by the shrinkage behaviour and can be discussed as follows:

If one tries to explain from fundamental issues in polymer science for natural and synthetic fibrous polymers, shrinkage is governed by the potential that it can occur (thermodynamic) and by the rate at which it can occur (kinetics). It is well established as per thermodynamics theory that there is a balance between cohesive energy and entropy when a process is at equilibrium. The cohesive energy between molecules retains the shape and dimension of a solid fibrous polymer while the entropy opens it and allows segmental relaxation that leads to shrinkage. This balance establishes the fibrous shape and stability that is disturbed and temporally fixed into a non-equilibrium position during manufacturing processes. When fibres swell in a liquid or are heated above their glass transition temperature during cleaning or drying in air, cohesive energy force weakens and entropy forces get strong. This relaxes the morphology and fibres shrink. But because polymeric fibers are viscoelastic, the thermodynamically feasible endpoints are not reached instantaneously. Under these conditions, the kinetics of the process will determine the dimensional properties of the fibres. Therefore, we can only delay relaxation shrinkage during textile cleaning but we cannot stop it.

The practical consequence is that relaxation shrinkage takes time and does occur cumulatively over several cleaning cycles. The phenomenon is well known as "progressive shrinkage". If one can find a cleaning and finishing process which delays perceivable relaxation shrinkage long enough to exceed a garment's life cycle, consumers will be satisfied. Non-aqueous cleaning used to do this, but it is much more difficult to manage with aqueous cleaning. When garment shrinks more than 2 or 3 percent. The garments do not fit well anymore and consumers will not like it. Shrinkage can occur during the washing, bleaching, drying, drycleaning or finishing processes. The new wet cleaning technology optimizes and controls the well-known process parameters: time, mechanical action, heat, and chemistry to reduce shrinkage. Shrinkage can be classified into two categories: felting and relaxation.

Felting shrinkage: This type of shrinkage is unique in wool because wool fibres have surface scales that cause differential friction effects. When wool fibres swell, as they do in water, the scales expand and are lifted. This increases differential friction between fibres and interlocks and compacts them, which causes felting shrinkage. It is possible to reduce but not eliminate the felting potential of wool with process additives that lower inter-fibre friction and reduce fibre swelling.

Relaxation shrinkage: During fabric and garment manufacturing, textiles are stretched, shaped and dried under tension. This causes latent stresses at the macroscopic level (between fibres and yarns) and at the microscopic level (within the fibre morphology). The macroscopic stresses are generally relaxed by mechanical action that allows movement between fibres and yarns. Microscopic stress is released by plasticisation. Plasticisation occurs when fibres swell in a liquid medium or when excessive energy (heat) is applied during drying. Either action lowers the cohesive energy between amorphous polymer segments and causes relaxation within the fibre matrix leading to shrinkage.

Visual and tactile perceptions are also vital from the point of view of consumers. Cleaning experts have the challenge to retain or restore the inherent properties that cause the desirable sensory attributes of textiles triggering positive purchasing decisions. Thus, it is important to retain the original colours, textures, and finishes during cleaning or to restore them if undesirable changes have occurred. Most drycleaners use fabric finishes to restore or improve the hand and feel of drycleaned fabrics. Fabric finishes for aqueous cleaning are also available to achieve the same desirable effects.

1.6   Concluding remarks

Accurate care labelling is necessary for the consumer to be able to adequately care for the garment or textile article and the mandatory labelling requirements make it imperative for manufacturers and retailers to ensure that goods manufactured and offered for sale comply with the requirements. The law requires it, and it also makes good business sense, as customers would patronise products that meet quality criteria and offer other tangibles such as good care advice. Claims in the care label in certain apparels that dye bleeding and staining can be prevented need to be verified as per the standard method. While it is possible to control selective colourant removal and staining, the complex and diverse nature and properties of colourants and textiles suggest that it is not easy to fulfill such a broad claim. The real issue here is proper dyeing and colour fastness evaluation during textile chemical processing. The coherent effort of textile and apparel manufacturers, retailers and textile care specialists is necessary to establish quality and test protocols to optimize garment performance for effective cleaning of their product.

1.7   References

1. ISO 3758 Textiles -- Care labelling code using symbols.
2. ASTM D3136 Standard Terminology Relating to Care Labeling for Apparel, Textile, Home Furnishing, and Leather Products.

3. ASTM D5489-01a Standard Guide for Care Symbols for Care Instructions on Textile Products.

4. FTC Care labelling guide, Clothes Captioning: Complying with the Care Labelling Rule, Available from: http://www.ftc.gov/bcp/edu/pubs/business/textile/bus50.pdf. [Accessed on 3rd March 2018.

5. 16 CFR Part 423 Trade Regulation Rule on Care Labeling of Textile Wearing Apparel and Certain Piece Goods, Available from:

http://www.ftc.gov/os/2000/07/carelabelingrule.htm.[Accessed on 4th March 2018]

2

NBC Protective finishes for defence applications

Dr. L. Ashok Kumar

Professor, Dept. of Electrical & Electronics Engineering,
PSG College of Technology, Coimbatore, Tamilnadu, India,
askipsg@gmail.com, Mob : 098432 81115

Abstract : An NBC (nuclear, biological, chemical) suit, also called a chemsuit or chem suit or chemical suit is a type of military personal protective equipment. NBC suits are designed to provide protection against direct contact with and contamination by radioactive, biological or chemical substances, and provide protection from contamination with radioactive materials and some types of radiation, depending on the design. They are generally designed to be worn for extended periods to allow the wearer to fight (or generally function) while under threat of or under actual nuclear, biological, or chemical attack. The civilian equivalent is the hazmat suit. The term NBC has been replaced by CBRN (chemical, biological, radiological, nuclear), with the addition of a new threat, radiological, meaning radiological weapon. NBC stands for nuclear, biological, and chemical. It is a term used in the armed forces and in health and safety, mostly in the context of weapons of mass destruction (WMD) clean-up in overseas conflict or protection of emergency services during the response to a terrorist attack, though there are civilian and common-use applications (such as recovery and clean up efforts after industrial accidents). In military operations, NBC suits are intended to be quickly donned over a soldier's uniform and can continuously protect the user for up to several days. Most are made of impermeable material such as rubber, but some incorporate a filter, allowing air, sweat and condensation to slowly pass through.

Keywords : NBC clothing, CBRN, NBC Weapons, Defence applications

2.1   Introduction

CBRN is weaponized or non-weaponized Chemical, Biological, Radiological and Nuclear materials that can cause great harm and pose significant threats in the hands of terrorists. Weaponized materials can be delivered using conventional bombs (e.g., pipe bombs), improved explosive materials (e.g., fuel oil-fertilizer mixture) and enhanced blast weapons (e.g., dirty bombs). Non-weaponized materials are traditionally referred to as Dangerous Goods (DG) or Hazardous Materials (HAZMAT) and can include contaminated food, livestock, and crops.

2.2   CBRN incident

An *accidental* CBRN incident is an event caused by human error or natural or technological reasons, such as spills, accidental releases, or leakages. These accidental incidents are usually referred to as DG or HAZMAT accidents. Outbreaks of infectious diseases, such as SARS, or pandemic influenza are examples of naturally occurring biological incidents.

An intentional CBRN incident includes:
- criminal acts such as the deliberate dumping or release of hazardous materials to avoid regulatory requirements
- the malicious, but non-politically motivated poisoning of one or more individuals
- terrorist acts (as defined in the Criminal Code of Canada and the Security Offences Act) that involve serious violence to persons or property for a political, religious or ideological purpose and/or that are a matter of national interest The response to an intentional CBRN incident may be similar to an accidental CBRN incident; however, intentional CBRN incidents differ because there are unique implications relating to federal/provincial/territorial responsibilities, public safety, public confidence, national security, and international relations.

CBRN incidents may include all or some of the following characteristics:
- Potential for mass casualties
- Potential for loss of life
- Potential for long term effects
- Creation of an extremely hazardous environment
- Relative ease and cheapness of production
- Initial ambiguity and/or delay in determining the type of material involved
- Potential use of a combination of CBRN materials each presenting different response requirements
- Narrow time frame in which to administer lifesaving interventions/ treatments;
- Need for immediate medical treatment for mass casualties
- Need for immediately available specialized pharmaceuticals
- Need for specialized detection equipment
- Need for timely, efficient and effective mass decontamination systems

- Need for organized, trained and equipped health service personnel to immediately augment local Fire-HAZMAT teams
- Need for pre-coordination within health services to establish medical treatment protocols, to stock pharmaceuticals and to determine treatment requirements
- Need to establish coordinated incident management/response procedures for such incidents;
- Need to ensure early warning systems for hospitals
- Need to establish early those who are affected and those at risk
- Need for active case finding versus passive case finding
- Need to work closely with Police on site and at health care facilities, as they perform their legal duties in relation to victim identification/registration and evidence gathering
- Need for a pro-active media policy to ensure the community is informed and thus its anxiety allayed.

Countermeasures include:
- Technical equipment such as respirators that can detect chemical agents and masks that prevent exposure
- Medical therapy and, for some agents, prophylaxis
- Organizational strategies, such as specially developed intelligence systems, standard operating procedures, and training
- Instruments of international law.

2.3   Biological Weapons

Biological weapons are weapons that achieve their intended effects by infecting people with disease-causing microorganisms and other replicative entities, including viruses, infectious nucleic acids, and prions. The chief characteristic of biological agents is their ability to multiply in a host over time. The disease they may cause is the result of the interaction between the biological agent, the host (including the host's genetic constitution, nutritional status and the immunological status of the host's population) and the environment (e.g., sanitation, temperature, water quality, population density). Biological agents are commonly classified according to their taxonomy (e.g., fungi, bacteria, viruses). This classification is important because of its implications for detection, identification, prophylaxis, and treatment. Biological agents can also be characterized by other features, such as infectivity, virulence, lethality, pathogenicity, incubation period, contagiousness and mechanisms of transmission, and stability – all of which affect their potential to be used as weapons. The following definition of these terms is based on those of J M Last, *A dictionary of epidemiology*, fourth edition, Oxford University Press, 2001.

- The **infectivity** of an agent is its capability to enter, survive and multiply in a host and may be expressed as the proportion of persons exposed to a given dose who become infected.

- **Virulence** is the relative severity of the disease caused by a microorganism (i.e., the ratio of clinical cases to the number of infected hosts). Different strains of the same microorganism may cause diseases of different severity. For example, infection due to *Brucella melitensis* is usually more severe than infection due to *B. suis* or to *B. abortus*.

- **Lethality** is the ability of an agent to cause death in an infected population. The case-fatality rate (i.e., the proportion of patients clinically recognized as having a specified disease who die as a result of that illness within a specified time) provides useful information on the clinical management of cases.

- **Pathogenicity** is the capacity of a microorganism to cause disease, and is measured by the ratio of the number of clinical cases to the number of exposed persons. The incubation period is the time between exposure to an infective agent and the first appearance of the signs and symptoms of the disease. The incubation period can be affected by many variables, such as the initial dose, virulence, route of entry, rate of replication, and the immunological status of the host.

- For those infections that are contagious, a measure of their **contagiousness** is the number of secondary cases following exposure to a primary case in relation to the total number of exposed susceptible secondary contacts. The mechanisms of transmission involved may be direct (i.e., direct contact between an infected and an uninfected person) or indirect. (i.e., through inanimate material that has become contaminated with the agent, such as soil, blood, bedding, clothes, surgical instruments, water, food or milk). Infections may also be through airborne droplets (i.e., through coughing or sneezing) or through vectors, such as biting insects. The distinction between types of transmission is important in selecting control measures. For example, direct transmission can be interrupted by appropriate handling of infected persons, while interrupting indirect transmission requires other approaches, such as adequate ventilation, chlorination of water, or vector control.

- **Stability** is the ability of the agent to survive the influence of environmental factors such as air pollution, sunlight and extreme temperatures or humidity.

2.4   Chemical weapons

Chemical weapons are those that are effective because of their toxicity: that is, their chemical action can cause death, permanent harm or temporary incapacity. Weapons that use chemicals as propellants, explosives, incendiaries or obscurants are not chemical weapons, even though the chemicals in them may also have toxic effects. Only weapons whose main goal is to have toxic effects are considered chemical weapons. Some toxic chemicals, such as phosgene, hydrogen cyanide and tear gas, may be used for both civil and peaceful, and hostile purposes. When they are used for hostile purposes, they are considered chemical weapons. A common way to classify chemical agents is according to the degree of effect (e.g., harassing, incapacitating or lethal).

- A **harassing agent** disables exposed people for as long as they remain exposed. They are acutely aware of discomfort, but usually able to remove themselves from exposure to it unless they are otherwise constrained. They usually recover fully a short time after exposure ends, and do not require medical treatment.
- An **incapacitating agent** also disables, but people exposed to it may not be aware of their predicament (e.g., certain psychotropic agents), or may be unable to function or move away from the exposed environment. The effect may be prolonged, but recovery may not require specialized medical aid.
- A **lethal agent** causes death for those exposed. This approach to classifying chemical agents is not particularly precise because the effects of chemical agents will depend on the dose received, and on the health and other factors that affect how susceptible people are to the agent. For example, tear gas is usually a harassing agent, but it can be lethal if a person is exposed to a large quantity in a small closed space. On the other hand, nerve agents are usually lethal, but may only incapacitate people who are exposed to a low concentration for a short time. If it is not possible to totally protect people from chemical weapons, protective measures should try to reduce their effect. For example, the use of pretreatment and antidotes in a nerve gas victim is unlikely to provide a complete "cure" but may reduce what would have been a lethal effect to an incapacitating one. Another form of classifying chemical agents is based on their effects on the body.

- **Nerve agents** gain access to the body usually through the skin or lungs, and cause systemic effects.
- **Respiratory agents** are inhaled and either cause damage to the lungs, or are absorbed there and cause systemic effects.
- **Blister agents** are absorbed through the skin, either damaging it (e.g., mustard gas) or gaining access to the body to cause systemic effects (e.g. nerve agents), or both. A further classification is based on the duration of the hazard.
- **Persistent agents** remain in the area where they are applied for long periods (sometimes up to a few weeks). They are generally substances of low volatility that contaminate surfaces and have the potential to damage the skin if they come into contact with it. A secondary danger is inhalation of any vapours that may be released. Persistent agents may consequently be used to create obstacles, contaminate strategic places or equipment, deny access to an area, or cause casualties. Protective footwear and/or dermal protective clothing will often be required in contaminated areas, usually with respiratory protection. Mustard gas and VX are persistent agents.
- **Non-persistent agents** are volatile substances that evaporate or disperse quickly, and may be used to cause casualties in an area that the group using the weapons wants to occupy soon after. Surfaces are generally not contaminated. The primary danger is from inhalation, secondary from skin exposure. Respirators are the main form of protection required. Protective clothing may not be necessary if concentrations are below skin toxicity levels. Hydrogen cyanide and phosgene are typical non-persistent agents. Chemical agents are often grouped according to their effect on the body, based on the primary organ system affected by exposure. Typical classes include:
- **nerve** agents or "gases" (e.g., sarin, VX, VR).
- **blood** gases or systemic agents (e.g., hydrogen cyanide).
- **vesicants** or skin blistering agents (e.g., mustard gas, lewisite).
- **lung** irritants, asphyxiants or choking agents (e.g., chlorine, phosgene).

2.5   Modeling the Chemical Protective Performance of NBC Clothing Material

The heat load, imposed by air-permeable NBC (nuclear, biological, and chemical)-protective suits, can be reduced by improving the air permeability of the suit. However, increased air permeability will reduce the chemical protective performance, since increasing the air permeability of the NBC-protective material will result in higher air velocities through the material. The NBC-protective clothing, currently in use by military forces, usually is an air permeable carbonbased garment. This clothing protects the wearer by adsorbing the hazardous chemical vapors onto the carbon. The thermal load of this type of clothing is low compared with that of impermeable clothing because of the relatively good transmission of air and water vapor. The heat stress on the wearer can nevertheless be a problem, especially in hot ambient environments. A reduction of the heat load can be achieved by improving the air permeability of the fabric. The flow through the material will then increase to a higher comfort can be achieved. However, increased air permeability will reduce the chemical's protective performance. To achieve a good compromise between comfort and protection, it is useful to understand the relationship between the chemical protective performance and the air permeability and thus the air velocity through the material.

A semi-empirical model is deduced that describes the breakthrough concentration of chemical vapor through NBC protective clothing material. The model describes the influence of the air flow through the material, the challenge concentration, and the type of agent on the chemical barrier properties of the fabric. By using the diffusion coefficient and the adsorption isotherm onto carbon, the effect of the type of agent onto the breakthrough concentration is also incorporated, but these dependencies were not validated here. A two-dimensional mathematical model was used to describe the dynamic behavior of the adsorber. The transport mechanism in the gas phase is described by a dispersed plug flow model accounting for mixing in the axial direction. The resistance for mass transfer between the bulk of the gas phase and the outer surface of the particle is calculated for an assumed thin film around the particle. Models that describe the breakthrough of vapour through carbon filters have different boundary conditions at the top and the bottom. Since the NBC-protective clothing consists of only one layer of carbon particles, this implies that there are actually different boundary conditions on the two sides of the carbon layer. This causes instabilities and therefore this model is not completely applicable. These boundary condition problems are solved by describing the breakthrough curve semi-empirically.

2.6   Theoretical

In theoretical analysis, only NBC-protective clothing materials of the carbon bead type have been taken into account. In this type of protective clothing, the chemical filtration is based on a single layer of small activated carbon spheres adhered to a carrier fabric (Figure 2.1).

When an activated carbon filter is challenged by a chemical agent vapor flow, the breakthrough curve of the effluent vapor concentration against time is typically S-shaped. Typical for carbon bead type fabrics is an initial step in the breakthrough curve. Immediately after exposure, a very small breakthrough concentration occurs that is roughly constant over a certain period of time (typically a few hours for NBC-clothing material). Another characteristic parameter of a breakthrough curve is the time at which 50% breakthrough occurs, which is determined by the adsorption capacity of the carbon. For both the initial breakthrough concentration and the 50% breakthrough time a model is developed. The total breakthrough curve is described as semi-empirically.

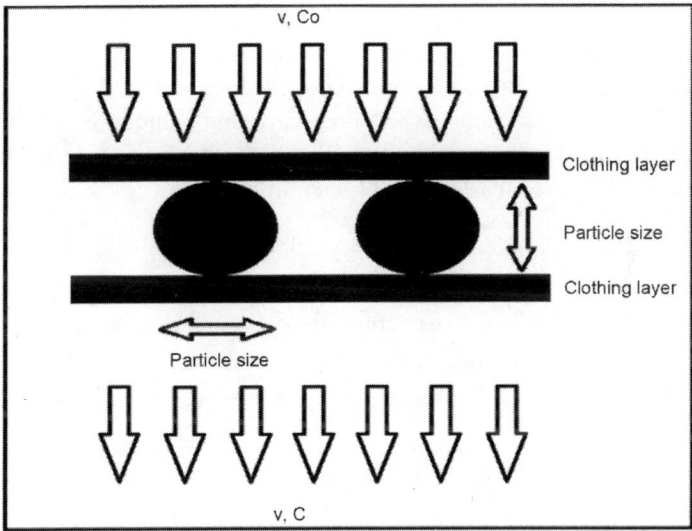

Figure 2.1: A schematic representation of a layer of NBC Clothing material

2.7   Optimizing the Protection against the Physiological Burden of CBRN Clothing

Soldiers can wear chemical, biological, radiological and nuclear (CBRN) protective clothing to be protected against warfare agents. The disadvantage of that clothing is that higher protection introduces higher physiological burden. Therefore, an optimum between comfort and protection must be found. Models of all relevant processes were created to find this optimum. The airflow profile around a cylinder with clothing representing a dressed human body part—was modelled. This flow profile was used for calculating the agent vapour breakthrough through the clothing and for calculating the deposition of agents onto the skin (as indicators for protection). The flow profile was also used for calculating the temperature profile around the body part and the relative humidity underneath and in the clothing (as representative for physiological burden). As a result, a tool was created, which can be used to identify the optimum properties of CBRN protective clothing, depending on the intended mission of the soldiers.

Soldiers can experience the threat of chemical warfare agents when working in the field. To protect themselves against these vapours or liquids, they can wear chemical, biological, radiological and nuclear (CBRN) protective clothing. The CBRN protective clothing, currently in use by military forces, is usually an air permeable carbon-based garment. The basis of the protection is adsorption of the toxic agent onto the carbon. The use of air-permeable materials in clothing reduces the thermal load offered by the clothing, by allowing air to flow through the material. However, at the same time, the airflow can transport the toxic agents through the clothing if the agent is not fully absorbed by the carbon. A way to identify the balance between physiological burden and protection is by modelling the important processes at hand and using these models to find the optimum. The system requirements following from this optimization will lead to requirements for subsystems and materials of the clothing. That way a simulation tool will be created, which can be used by several different types of users, e.g.,

- military planners for identifying the ideal suit for a specific mission;
- producers of CBRN clothing for creating the optimum clothing design;
- researchers for finding the ideal balance between physiological burden and protection;
- quality testing for identifying key parameters.

To model the protective performance of air permeable CBRN suits, several scales in the model have to be considered. Figure 1 illustrates the three scales:

- microscale: the protective performance of the material itself;
- mesoscale: the effect of air flow around body parts on the deposition on the skin;
- macroscale: the model of the whole system (suit).

To model all the processes which occur underneath clothing in full detail, computational fluid dynamics (CFD) is required. For instance, the modelling of the airflow distribution around cylinders was described by using CFD.

Previously the micro-scale processes with respect to protection were studied. Furthermore, parameters like the air velocity distribution, the concentration distribution, and the temperature distribution were analyzed previously. Several types of turbulence models were examined.

Figure 2.2 shows an example of the results of the very time-consuming DNS (direct numerical simulation) calculations. A three-dimensional flow profile around a cylinder is shown. Behind the cylinder, the (oscillating) wake of the airflow can be seen. This figure is a snapshot of a 16-s movie. It is convenient to have a model, which can predict the behaviour of these processes without using CFD, without high loss of accuracy. This is so, because of integration with other models; ease of use; ease of changing parameters; calculation time (when, e.g., DNS is used); etc. Therefore, all these processes have to be modelled with basic one-dimensional physical transport equations.

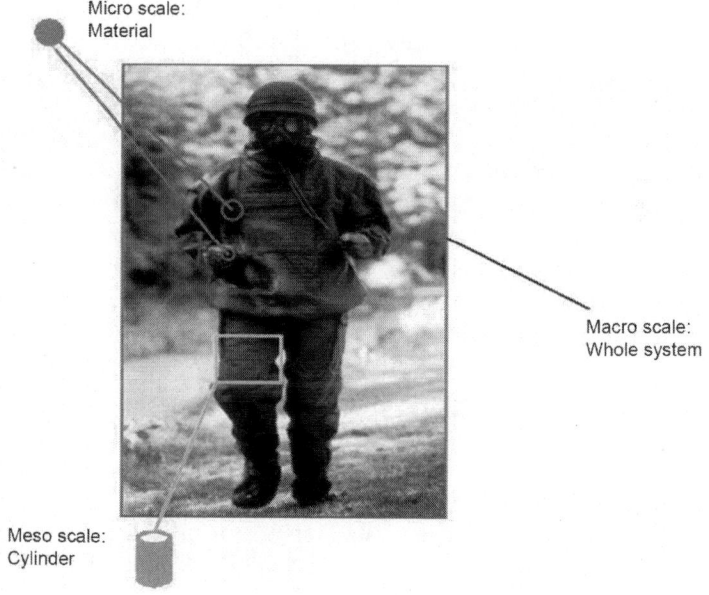

Figure 2.2: The process scales, important for the protection of a chemical, biological, radiological and nuclear (CBRN) suit.

2.8   Processes

Several processes occur when protective clothing is worn. To investigate both protective performance and comfort parameters, the following processes can be identified:

- airflow around, through and underneath the clothing;
- agent vapour breakthrough through the clothing;
- agent vapour concentration underneath the clothing;
- agent deposition onto the skin;
- water vapour penetration through the clothing;
- water vapour concentration underneath the clothing;
- evaporation and condensation of water from/in the clothing and from/ onto the skin;
- temperature distribution through the clothing, underneath the clothing;
- skin temperature and core temperature;
- adsorption and absorption processes for agent and water in the clothing.

2.9   Agent vapour breakthrough through clothing

If the flow distribution around, through and underneath the clothing is known, a concentration distribution can be derived. The first step will be to calculate the agent vapour breakthrough through the protective clothing. A model was derived earlier, which describes the vapour breakthrough through CBRN clothing as a function of time. Only CBRN protective clothing materials of the carbon bead type have been taken into account. In this type of protective clothing, chemical filtration is based on a single layer of small activated carbon spheres adhered to a carrier fabric. When an activated carbon filter is challenged by a chemical agent vapour flow, the breakthrough curve of the effluent concentration against time is typically S-shaped. Typical for carbon bead type fabrics is an initial step in the breakthrough curve. Immediately after exposure, a very small breakthrough concentration occurs that is roughly constant over a certain period of time (typically a few hours for CBRN clothing material).

2.10   Air flow model

A model was developed earlier, which describes pressure distribution and velocity through and underneath the clothing around the cylinder. To model airflow distribution around a human limb, simplification has been made by assuming this to be a two-dimensional problem. The body part is represented by a cylinder which is placed in airflow. At the front of the (dressed) cylinder, the wind will penetrate the clothing and at the back, it will flow back to the environment. If only the homogeneous perpendicular flow of the outside wind is assumed and the airflow underneath the clothing is modelled only one-dimensionally, the process can be solved without using CFD.

Figure 2.3 shows a schematic picture of the set-up. The wind is blowing against the clothing around the cylinder with a velocity $v0$. At the front of the cylinder, at point C (the stagnation point) the wind will partly flow through the porous clothing and will continue to flow underneath the clothing. At different points around the cylinder (different angles from the stagnation point, e.g., A), the velocity through the clothing will be different, because the pressure around the cylinder varies. Up to a certain angle, the wind will flow into the air gap, but at larger angles, the wind will flow out again. This process depends on the pressure.

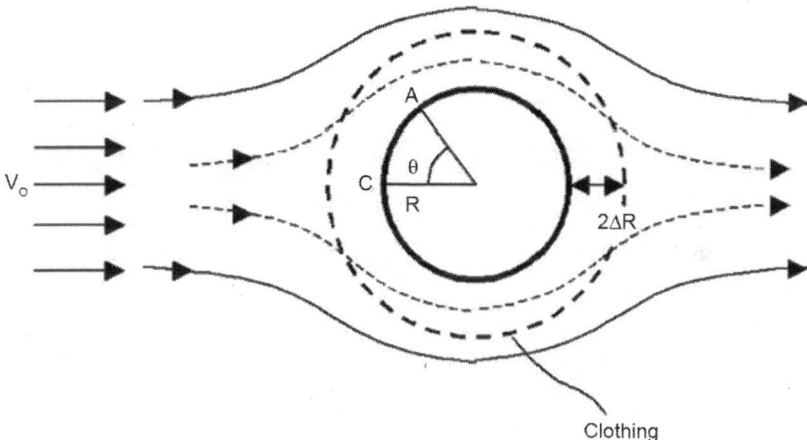

Figure 2.3: A schematic representation of a body part dressed in clothing, with airflow around and underneath the clothing.
Notes: A, C—stagnation points; cf. p. 167–8 for definitions of the other symbols.

2.11   Agent vapour concentration and deposition model

The breakthrough concentration model and the airflow distribution model can be used as input in the next step to model the concentration distribution underneath the clothing and the local mass deposition on the cylindrical surface. A model was developed earlier, which describes the vapour concentration underneath the clothing and the deposition onto the cylinder [10]. For this step, it is assumed that (a) all the mass of the chemical agent which reaches the cylinder surfaće is adsorbed, and (b) the mass transport to the surface is described by the penetration theory. The first assumption implies that the concentration on the surface of the cylinder is always zero. A continuity equation (a mass balance for a slice of air underneath the clothing) describes the agent vapour concentration underneath the clothing as a function of the angle θ with respect to the wind direction around the cylinder (Figure 2.4).

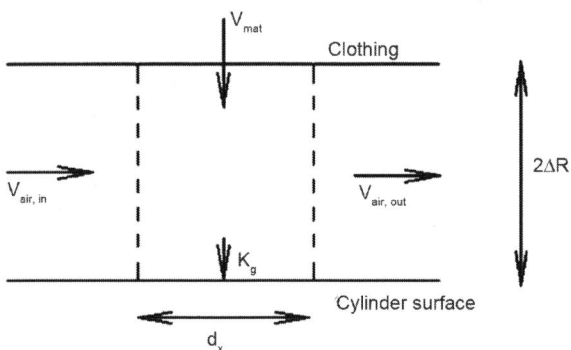

Figure 2.4: A volume element of air between the clothing and the surface of the cylinder. *Notes:* dx—tangential step along the surface of the cylinder (m); cf. p. 167–8 for definitions of the other symbols.

2.12   Moisture and temperature profile model

The water vapour concentration can be described similarly to the agent vapour concentration (Figure 2.5). However, some adaptations have to be made onto the model. The main differences are (a) people sweat, thus the cylinder produces water vapour; and (b) water vapour can be absorbed in the clothing and can even condensate in the clothing or onto the skin, whereas the agent vapour adsorbs onto the adsorber (usually carbon). A continuity equation (a mass balance for a slice of air underneath the clothing) describes the water vapour concentration underneath the clothing as a function of the angle θ with respect to the wind direction around the cylinder. A model was developed, which describes the humidity profile and temperature underneath the clothing.

A modelling tool is being developed; it can be used to optimize the physiological burden effects and protective performance of CBRN clothing around a cylinder. To identify the most adequate type of CBRN protective clothing, a compromise between physiological burden and protection has to be found. The current model focuses on temperature, RH, air velocity and agent concentration underneath clothing around a cylinder. As expected, better protection can only be created when a higher physiological burden is also tolerated. This means a different physiological burden is allowed for every mission, thus the protection will be mission dependent. The current models do not yet include leakage of the agent through, e.g., closures. When the models are extended toward a whole system, they can be used to create an optimum design of the suit. Closures and other leakages of agent will influence the concentration underneath the clothing and thus the total deposition onto the skin (and thereby the protection). Induced air pumping (or the bellowing effect) will affect the air velocity underneath the clothing and thus the physiological burden. When these aspects are modelled as well, a better optimum can be found. Another missing part in the model is a body model, combined with a good sweat model. In the current model, the body core temperature and the sweat rate are kept constant. This is not yet very realistic. Therefore, it is advisable to include such a model into the package. This is especially important when the whole system model is built.

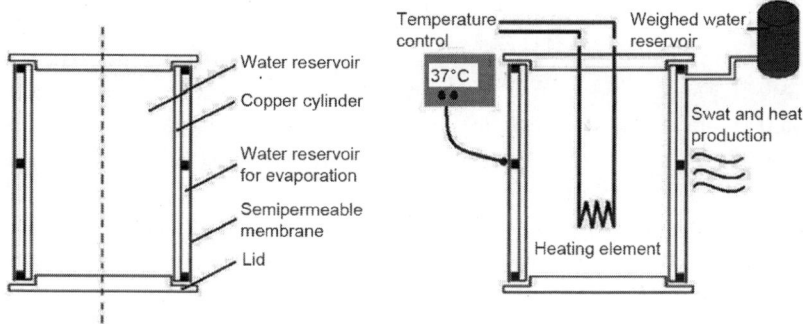

Figure 2.5: A schematic representation of the experimental set-up.

2.13   Influence of hydration volume and ambient temperature on physiological responses while wearing CBRN protective clothing

Emergency service personnel are required to wear specialized personal protective equipment (PPE) when conducting operations in environments contaminated with chemical, biological, radiological and/or nuclear (CBRN) materials. However, the PPE (by necessity) typically has limited or no air permeability, which greatly impedes evaporative heat loss and increases thermal strain, placing personnel at greater risk of heat injury (for review, see Cheung et al. 2000). Ensuring that personnel receives adequate hydration before, during and after deployment to offset fluid losses through sweating is one strategy to reduce the extent of dehydration. However, despite sweat rates being greater in a hot environment, at any given level of thermal strain, individual variation in sweat rates is high even among those with similar levels of aerobic fitness and states of heat acclimation. Thus, recommending a single rate or volume of fluid replacement to offset rates of fluid loss is potentially difficult.

Nevertheless, it is commonly assumed that body mass loss should not exceed 2% as these levels are consistently associated with increased thermal strain, cognitive impairment and decreased work or exercise performance. Interestingly, levels of dehydration as small as 1% of body mass have been reported to reduce tolerable limits of thermal strain and work productivity in firefighters wearing standard issue firefighter PPE. Thus, ensuring proper rehydration schedules during deployments in PPE can safely increase levels of thermal strain achieved and reduce the individual risk of becoming a heat casualty. Further, individuals who are not regularly active may experience greater impairment with smaller volumes of fluid loss due to their lower plasma volumes and their inability to maintain gut barrier integrity (i.e. the proximity of endothelial cell junctions) during exertional heat stress. Thus, for emergency service personnel, who may not be the equivalent of the endurance athlete, there is a need to provide guidelines for hydration prior to and during deployments that limit the level of dehydration to less than 1% of body mass. This may optimize work productivity and decrease the risk of exertional heat injury. Due to the restriction that the CBRN PPE confers on heat transfer from the body, variations in the external ambient temperature have less influence on thermal strain when rates of heat production are high and deployment times are less than 1 h. Thus, adjustments to rehydration guidelines due to ambient temperature may only be necessary with lower rates of heat production and longer heat stress exposures in the CBRN environment. UK Emergency Service personnel operating in a CBRN environment may be deployed in one of three currently available PPE. First, the gas-tight suit (GTS), which incorporates extended duration breathing apparatus and does not permit fluid replacement during deployments. However, despite this restriction, the ingestion of a bolus of fluid prior to a CBRN deployment to prevent excessive dehydration has not been investigated.

Second, the powered respirator protective suit (PRPS), which permits the use of a back-mounted water delivery system, and finally the civil responder mark 1 (CR1) suit, which also incorporates a respirator and permits the use of a back-mounted water delivery system. Emergency service personnel are encouraged to drink ad libitum during and following deployments. In particular, when deployed in the CR1, personnel are instructed to drink every 20 min but specific guidelines that consider both the work rate and environmental temperature do not exist. In addition to promoting ad libitum fluid replacement, personnel are advised to consume standard volumes of fluid at various time intervals that vary depending on the intensity and duration of the activity. However, these guidelines are no adjusted to the environmental conditions or to body mass. Therefore, it is evident that fluid replacement guidelines for all of the UK emergency services operating in a CBRN environment could be improved with strategies developed from evidence-based findings specific for the PPE worn. Thus, the purpose of this study was to examine whether low (L; 5 ml/kg per h) or high (H; 10 ml/kg per h) rates of fluid replacement were appropriate in a moderate (188C) or hot (308C) environment to prevent fluid losses from exceeding 1% of body mass for emergency service personnel wearing different CBRN PE. These rates of fluid replacement were chosen based on comparable rates of fluid loss reported for exercise in CBRN PPE in similar environmental conditions (McLellan et al. 1993, Wilkinson et al. 2009). The outcomes will contribute to the development of guidelines to assist the emergency services in producing fluid replacement strategies for personnel deployed wearing CBRN PPE.

2.14   Heat stress in chemical protective clothing: porosity and vapour resistance

One of the major problems associated with chemical warfare (CW) protective clothing (nuclear, biological and chemical (NBC) or chemical, biological, radiological and nuclear (CBRN)) and their industrial chemical protective counterparts is the additional heat load created by these garments. For breathable CW over-garments, traditionally containing an activated carbon/charcoal layer in a foam or other structure with a liquid repellent/spreading textile on top (bi-pack), research on alleviating heat stress has been focused in two directions. The first is reducing material thickness and thus heat and vapour resistance; the second is the use of thin, selectively permeable breathable membranes (the latter refers to waterproof membranes that would stop any CW agents (gas or liquid) from penetrating, but would allow the passage of moisture vapour. The word 'breathable' is often used for 'vapour permeable' in this context and does not relate to permeability for breathing gases such as O2, CO2 or N2. Even though the thickness and the heat and vapour resistance of CW-protective clothing materials have been reduced by over 50% over the past decades, this has not necessarily resulted in major improvements in terms of heat strain for the wearer. The cause for this lack of observable effect is that the thickness and heat/vapour resistance of the complete clothing assembly (the over-garments are usually worn over underwear and standard combat clothing) is not only determined by the outer material layers, but is a factor of all clothing layers, including all enclosed and adjacent air layers. This total thickness, as well as the total heat and vapour resistance, is hardly affected by a reduction in thickness of a single layer of the package. A different approach, which so far has received only minimal attention in a CBRN context, is the increase in air permeability of the garment materials. Such an increase could result in the improved ventilation of the clothing microclimate (especially during movement and in wind) and thus lead to reduced heat and vapour resistance.

An obvious fear is the reduction in chemical protection, concomitant with increasing air permeability. Thus, the first question to be answered is whether air permeability of common CW protective garments can be improved, while sustaining appropriate levels of chemical protection. Research on the effect of increased air permeability on CW-protection by TNO Defense, Security and Safety have shown that, for some types of CW-materials, protection can be maintained at higher levels of air permeability. This opened the floor for investigations on the relationship between air permeability and clothing heat and vapour resistance for CBRN clothing. In parallel to the trend for thinner, activated charcoal-loaded, over-garments the concept of selectively permeable membranes as the outer layer was introduced, which should provide a higher level of liquid protection. Finally, there is the combination of semipermeable membranes (liquid repellent, but permeable for water vapour and possibly vaporized CW agents) with a charcoal layer underneath, increasing the liquid protection compared to the air permeable bi-pack systems. Some of these membrane-based ensembles of these latter two concepts have vapour resistances similar in magnitude to the textile-based systems and, hence, one would expect heat stress in these garments to be similar to that in the textile versions, but with the benefit of higher protection.

Recently, researchers studied the effect of porosity of protective garments using the approach of measuring critical heat stress wet bulb globe temperature (WBGT) limits for such garments or measuring critical metabolic rate limits. They showed that these criteria correlated better with air permeability (convective permeability) than with vapour resistance of the fabric (diffusive permeability, e.g. measured as moisture vapour transfer rate). They further suggested that there was an upper limit to the porosity, above which no extra benefit for heat strain was observed. In their studies, they used lightweight commercially available coveralls as the outer layer, of which, in some, the air permeability was altered by puncturing them indifferent densities. Literature shows that the ventilation of the garments' microclimate in commercial protective coveralls that were also punctured with a grid of fine needles in different densities, thereby creating test garments of identical material, stiffness and fit, differing only in air permeability. They observed an increase in total ventilation with increasing air permeability, both for passive and active subjects. Ventilation through openings (wrist, neck, ankles) decreased, however, with increasing air permeability, probably due to a reduced pressure fluctuation in the more air-permeable garments. All garments used in the mentioned studies were civil designs and are much lighter and more flexible than typical military chemical protective garments.

The present study was performed with two goals. The first was to compare two CW protection concepts in terms of heat stress: the more traditional military air-permeable CW garment to an air-impermeable, membrane-based garment of higher liquid protection level, i.e. two ensembles with identical inner layers, minimally different total static vapour resistances, but mainly differing in porosities. The second was to compare the number of charcoal, foam-based CW protective garments of identical stiffness, design, fit, etc., differing only in porosity and to determine whether a relevant physiological heat strain difference between these garments is observed. For the latter set of garments, measurements of microclimate ventilation with tracer gases were also performed. When comparing ensembles with a textile vs. a membrane outer layer, which have similar static heat and vapour resistances measured for the sum of fabric samples, a higher heat strain is observed in the membrane ensemble. This is because in actual wear the air permeability of the textile version improves ventilation and allows better cooling by sweat evaporation. For garments with identical thickness and static dry heat resistance, but differing levels of air permeability, a strong correlation of microclimate ventilation due to wind and movement with air permeability is present. This is reflected in lower values of core and skin temperatures and HR for garments with higher air permeability. For HR and core temperature, the two lowest and the two highest air permeabilities form two distinct groups, but do not differ within the group. Based on protection requirements it is concluded that air permeability increases can improve heat strain levels, allowing optimization of protective clothing. The effect of increasing air permeability may be stronger than improvements through the reduction of the material thickness.

2.15   Methods of evaluating protective clothing relative to heat and cold stress: thermal manikin, biomedical modeling, and human testing

Personal protective equipment (PPE) refers to clothing and equipment designed to protect individuals from chemical, biological, radiological, nuclear, and explosive hazards. The materials used to provide this protection may exacerbate thermal strain by limiting heat and water vapor transfer. Any new PPE must, therefore, be evaluated to ensure that it poses no greater thermal strain than the current standard for the same level of hazard protection.

One of the most serious health hazards related to personal protective equipment (PPE) is thermal stress. PPE is designed to protect against natural or human-made hazards, such as chemical, biological, radiological, nuclear, and explosive hazards, and includes firefighting, first responder, HAZMAT, and certain occupational and combat uniforms. (1) PPE ranges from garments designed to protect from spills, to fully encapsulated protective suits with rebreathing apparatus, to the inclusion of body armor and helmets. The combination of PPE characteristics, environmental conditions, and required physical activity can result in thermal stress. Heat stress is elevated by multiple factors when wearing PPE. Added weight, bulk, and a layering effect of the addition of PPE all increase metabolic demands. (2) PPE also increases insulation and reduces water vapor permeability, both of which limit the ability to dissipate heat and evaporate water vapor from sweat. (3) This can increase body core temperature and limit work intensity or duration. During cold exposure, heat stress can occur if PPE is worn in combination with extreme cold weather clothing while working at a high metabolic rate; however, during rest periods, excessive heat loss can occur, particularly if clothing has become wet from sweat or precipitation. (4,5) In the cold, extremity (hand, foot) temperatures, rather than body core temperature, may be the critical factor for health and performance. (5) All new clothing and equipment for the military undergoes a health hazard assessment that evaluates health and performance risks associated with various hazards, including temperature extremes of heat and cold. (6) The physiologists, biophysicists, and physicians at the U.S. Army Research Institute of Environmental Medicine (USARIEM), in concert with the industrial hygienists at the U.S. Army Public Health Command, use a multidisciplinary approach in evaluating PPE to ensure the health and safety of individuals working in thermally stressful environments. This systematic evaluation includes biophysical evaluations to determine insulation and permeability of textiles on a guarded hot plate and ensembles on a thermal manikin; biomedical modeling to predict thermal strain, including work limits and fluid requirements; and human testing to quantify physiological responses in a controlled laboratory setting that simulates the conditions under which the PPE will be used if fielded.

This three-tiered approach provides much more information than any single test method and is a critical part of health hazard assessments. Typically, development of new PPE focuses on improving the level of protection from external hazards. This must be done without compromising thermal strain that can limit work unless it is determined that the benefit of increased protection outweighs the impact of added thermal stress. This review outlines the process used by the U.S. Army to evaluate new PPE for protecting health and performance with respect to temperature extremes. If this is considered early in the development process, health hazards related to temperature extremes can be mitigated while maintaining or improving the effectiveness of the PPE for protection from external hazards. While some examples of PPE presented in this review are specifically for military use, other applications include Homeland Security, law enforcement, and firefighting. Military PPE is sometimes adapted for industry use, and commercial PPE may be adapted for military use if it meets the criteria. Thus, a wide range of clothing developers could benefit from understanding the process of evaluation of thermal stress where the goal is to allow workers to be more productive for longer periods when working in temperature extremes.

The protective clothing used in the World War I against chemical warfare agents (CWA) consisted of rubber clothing, which, together with gloves and boots, to cover the entire body apart from that protected by the mask. Clothing of this kind is usually characterized as impermeable. This means that CW agents cannot pass through the material and also the fact that perspiration released from the skin is also prevented from passing out. Consequently, to wear clothing of this kind for longer periods is extremely uncomfortable and in hot climates the period during which protective clothing of this kind can be worn will be very short. In order to reduce the heat load developed due to wearing of impermeable protective clothing, permeable clothing was designed and developed, where a layer of finely distributed active carbon, either bound in polyurethane foam or as particles of carbon, is bound between two layers of textile. A layer of this kind consisting of active carbon which permits water vapour released from the body to pass through. The active carbon absorbs CWA and thereby prevents them from passing through to the skin. This layer of carbon is never used alone but is combined with different textiles.

A present CWA protective suit is an example of clothing made of permeable material. The largest difference is that inside the impregnated outer material there is a layer of active carbon on a suitable carrier. The CWA protective suit can be used instead of a battle dress or as an overall placed over the uniform. An alternative is to use inner clothing with a layer of carbon which is worn underneath the normal uniform. Impermeable suits can be used in severely contaminated environments, e.g., during decontamination. The heat load can be reduced by ventilating the clothing with fans. However, this solution is too vulnerable to be used. In order to achieve short-term CWA protection, it is possible to use overalls made of different plastic material. Many research and development occurred around the globe to achieve the body protection equipment time to time and are discussed sequentially thought this chapter.

2.16   References

1. W.B. Marzi, "Development of a new impermeable NBC protective suit for German civil defence" in the proceedings of the third international symposium on protection against chemical warfare agents, Stockholm, Sweden 11-16 June 1989, pp. 21-24, Swedish Defence Research Establishment, UMEA.

2. M.G. Katz, "A new approach to heat stress relief in chemical protective clothing" in the proceedings of the third international symposium on protection against chemical warfare agents Stockholm, Sweden 11-16 June 1989, pp. 25-31, Swedish Defence Research Establishment, UMEA.

3. G. Huzhang, H. Qitai and Z. Lei, "New carbon containing protective fabrics" in the proceedings of the third international symposium on protection against chemical warfare agents Stockholm, Sweden 11-16 June 1989, pp. 33-38, Swedish Defence Research Establishment, UMEA.

4. G. Fang, H. Dingmao, L. Guojin, Z. Weixin, and L. Jiangge, "Adsorption properties of carbon containing flannel" in the proceedings of the third international symposium on protection against chemical warfare agents Stockholm, Sweden 11-16 June 1989, pp. 39-44, Swedish Defence Research Establishment, UMEA.

5. G. Fang, "Chinese permeable chemical protective suits. Its test and evaluation" in the proceedings of the second international symposium on protection against chemical warfare agents, Stockholm, Sweden 15-19 June 1986, pp. 51-58, National Defence Research Establishment, Umea, Sweden.

6. E.E. Alexandroff, "Saratoga carbon pellet technology in chemical warfare protective clothing" in the proceedings of the second international symposium on protection against chemical warfare agents, Stockholm, Sweden 15-19 June 1986, pp. 67-76, National Defence Research Establishment, Umea, Sweden.

7. E.E. Alexandroff, "PBI Saratoga new and improved CWU/66p chemical protective clothing system for aircrew application" in the proceedings of the third international symposium on protection against chemical warfare agents, Stockholm, Sweden 11-16 June 1989, pp. 63-70, Swedish Defence Research Establishment, UMEA.

8. E. Helper, "The new personal chemical protective suit of the U.S military services" in the proceedings supplement of the sixth international symposium on protection against chemical warfare agents", Stockholm, Sweden May 10-15, 1998, pp. 200, Swedish Defence Research Establishment, UMEA.

9. S.G. Maroldo, "Carbonaceous resins-sorbents for chemical protection" in the proceedings of the third international symposium on protection against chemical warfare agents, Stockholm, Sweden 11-16 June 1989, pp. 71-78, Swedish Defence Research Establishment, UMEA.

10. E. Helper, "The concept of using and under garment for the NBC protection" in the proceedings of the fourth international symposium on protection against chemical warfare agents, Stockholm, Sweden 8- 12 June 1992, pp. 59, National Defence Research Establishment, Umea, Sweden.

11. T. Stoll, "New generation of permeable NBC protection clothing for hot climate conditions" in the proceedings supplement of the sixth international symposium on protection against chemical warfare agents, Stockholm, Sweden May 10-15, 1998, pp. 202, Swedish Defence Research Establishment, UMEA.

12. V. Obsel and V. Stein, "Czechosolavak textile adsorbents and possibilities of their utilization for noxious compounds trapping" in the proceedings of the fourth international symposium on protection against chemical warfare agents, Stockholm, Sweden 8-12 June 1992, pp. 63, National Defence Research Establishment, Umea, Sweden.

13. P. Nousiainen, M. Nieminen, A. Vuori, and M. Ranta, "Visac activated carbon fibers in composites for chemically protective clothing" in the proceedings of the fourth international symposium on protection against chemical warfare agents, Stockholm, Sweden 8-12 June 1992, pp. 61, National Defnce Research Establishment, Umea, Sweden.

[14] W. Vogel, "Cross laminated NBC protective films" in the proceedings of the third international symposium on protection against chemical warfare agents, Stockholm, Sweden 11-16 June 1989, pp. 55-60, Swedish Defence Research Establishment, UMEA.

15. J.F. Stampfer and R.J. Beckman, "A screening test for selecting chemical protective clothing" in the proceedings of the second international symposium on protection against chemical warfare agents, Stockholm, Sweden 15-19 June 1986, pp. 41-49, National Defence Research Establishment, Umea, Sweden.

16. B.J. Davey, "Degradation of human performance with use of chemical protective clothing" in the proceedings of the fourth international symposium on protection against chemical warfare agents, Stockholm, Sweden 8-12 June 1992, pp. 371-372, National Defence Research Establishment, Umea, Sweden.

17. K.W. Ang, F.K. Lee and P.S. Quek, "New protective material test system: Evaluation of multiple challenge chemicals" in the proceedings supplement of the sixth international symposium on protection against chemical warfare agents, Stockholm, Sweden May 10-15, 1998, pp. 197, Swedish Defence Research Establishment, UMEA.

18. R.J.V. Eenennaam, J. Kaaijk, M. Leeuw, H.F.G Oudmayer and G.J. Woudenberg, "Automated test system for NBC protective material" in the proceedings supplement of the sixth International symposium on protection against chemical warfare agents, Stockholm, Sweden May 10-15, 1998, pp. 198, Swedish Defence Research Establishment, UMEA.

19. J. Medema, P.T. Van Raaij and P.M.M. Wittgen, "Protection afforded by NBCF suits (influence of wear and tear" in the proceedings of the second international symposium on protection against chemical warfare agents, Stockholm, Sweden 15-19 June 1986, pp. 77-89, National Defence Research Establishment, Umea, Sweden.

20. G.R. Magin, "Decontamination of over garments" in the proceedings of the second international symposium on protection against chemical warfare agents, Stockholm, Sweden 15-19 June 1986, pp. 37-40, National Defence Research Establishment, Umea, Sweden.

21. J. Medema and P.M.M. Wittgen, "Effect of laundering on antichemical protective clothing" in the proceedings of the fourth international symposium on protection against chemical warfare agents, Stockholm, Sweden 8-12 June 1992, pp. 65-73, National Defence Research Establishment, Umea, Sweden.

22. S. Thandavamoorthy, N. Gopinath and S. S. Ramkumar, "Selfassembled honeycomb polyurethane nanofibers" *J. Appl. Poly. Sci.*, vol. 101, pp. 3121-3124, September 2006.

23. S. Ramakrishna, K. Fujihara, W.E. Teo, T. Yong, Z. Ma and R. Ramaseshan, "Electrospun nanofibers: Solving global issues" *Mater. Today*, vol. 9, pp. 40-50, March 2006.

24. R. Ramaseshan, S. Sundarrajan, Y. Liu, R.S. Barhate, N. L. Lala and S. Ramakrishna, "Functionalized polymer nanofibre membranes for protection from chemical warfare stimulants" *Nanotechnology*, vol. 17, pp. 2947-2953, June 2006.

25. M. Boopathi, M.V.S. Suryanarayana , A.K. Nigam, P. Pandey, K. Ganesan, B. Singh and K. Sekhar, "Plastic antibody for the recognition of chemical warfare agent sulphur mustard", *Biosens. Bioelectron.*, vol. 21, pp. 2339-2344, June 2006.

26. A. Baghel, M. Boopathi, B. Singh, P. Pandey, T.H. Mahato, P.K.Gutch, K. Sekhar, "Synthesis and characterization of metal ion imprinted nano-porous polymer for the selective recognition ofcopper, *Biosens. Bioelectron.*, vol. 22, pp.3326-3334, June 2007.

27. I. Luzinov, P. Brown and S. Husson, "Molecularly imprinted fibers with recognition capability" American National Textile Center Research Briefs: June 2007 [on line] available: www.ntcresearch.org/ pdf-rpts/Bref0607/Briefs07-TOC.pdf [Accessed September 4, 2008].

3

Denim preparation and finishing

Dr. A. K. Choudhary and Sheena Bansal

Department of Textile Technology, NIT, Jalandhar

Abstract : Among all the textile products, no other fabric has received such a wide acceptance as denim. Denim has always been the perfect blank canvas to express your personal style. Everyone has a go-to silhouette that they make their own. But while you mindlessly head to the boot cut, skinny, or high-waist section at your favourite denim shop, it means more than you think. And it's not just the silhouette you choose, it's the way the style them that really counts. The magic only really happens when the colour is faded and the jeans are properly worn in. It's a thing of beauty, really. Over time, a completely individual look develops. There is a wide variety of different denim fabrics like vintage denim, black denim, khadi Denim, stretch denim and many others.As there are different fabrics and there are different washes given to create personified look like bleached wash, dark wash, stone wash and many others. Manufacturers also utilized the situation by coming up with innovative designs like new colours, embroidery and patch work. Denim washing remains the most popular among manufacturer as well the customer. Denim finishing treatments refer to both mechanical and chemical processes for achieving diverse distressed or abraded looks and any number of distinctive designs to fit consumers' requests. The whole idea of finishing is about giving your fabrics the final touch to create the right high-quality fashion style.

Keywords : Denim, Yarns, Fabrics, Coloration, Finishing and Ranges

Among all the textile products, no other fabric has received such wide acceptance as denim[1]. Denim has always been the perfect blank canvas to express your personal style. Everyone has a go-to silhouette that they make their own. But while you mindlessly head to the boot cut, skinny, or high-waist section at your favourite denim shop, it means more than you think. And it's not just the silhouette you choose, it's the way the style them that really counts[2]. The magic only really happens when the colour is faded and the jeans are properly worn in. It's a thing of beauty, really. Over time, a completely individual look develops[3]. There is a wide variety of different denim fabrics like vintage denim, black denim, khadi denim, stretch denim and many others. As there are different fabrics and there are different washes given to create a personified look like a bleached wash, dark wash, stone wash, and many others. Manufacturers also utilized the situation by coming up with innovative designs like new colours, embroidery, and patchwork[4]. Denim washing remains the most popular among the manufacturer as well as the customer. Denim finishing treatments refer to both mechanical and chemical processes for achieving diversely distressed or abraded looks and any number of distinctive designs to fit consumers' requests. The whole idea of finishing is about giving your fabrics the final touch to create the right high-quality fashion style.

3.1   Introduction

Denim fabric is coarser than as usually woven fabric. Generally, coarse yarn is used for making denim. After yarn dyeing and weaving, denim fabric is sent to the finishing section or the washing section for adding some properties to the fabric. Denim washes are basically of two types, Mechanical Wash and Chemical Wash. While there are many different processes used to finish fabrics, the most common ones used on denim include singeing, scouring, bleaching, desizing, mercerizing, sanforizing, and bio-polishing, each designed to accomplish a specific goal. Besides giving these essential finishes there are some additional finishes which can be given to change the look of the fabric[5]. Getting the unique look with a superior reduction of back-staining is fast and easy, with various wet and dry finishing techniques such as washing, ripping, batching, crushing, whiskers, sandblasting, screen-printing, tinting, and bleaching. All these techniques give the fabric a softer feel, enhance its appeal without any strength loss, and improve wear life, color retention, and fashionable flair. Special finishing processes and garment enhancements that include the use of new technology and chemicals are all aimed at new possible effects of fabric look and originality, to satisfy customers' demands and increase the potential of the denim market. A range of denim treatment methods is used, such as diverse types of washing, dyeing, and bleaching, in order to get the desired effects in terms of look and feel[6].

3.2   Role of yarn in denim finishing

Cotton and its intimate blends with other cellulose-based fibers is a material used for most denim-type fabrics. Currently, it is possible to obtain different types of thread in the denim yarn market, according to the manufacturing system used: Rotor or Open-End (OE) System, Ring System or Compact System. The type of yarn used in the construction of fabric and later in garment production decidedly influences the latter's final appearance. The use of yarn that displays certain irregularities is related to the current tendency to produce final garments with worn-out or 'old' appearance. During warp dyeing, the quality of the yarn, as well as its type, is an important factor when it comes to enduring tensions with which the ranges work. Generally, open-end type yarn has a good absorption capacity in short periods of bath-material contact in dye vats. Ring type yarn also has a good absorption capacity, although lower than that of open-end yarn. If the ring yarn is additionally combed (with a greater orientation of fibers in the same direction), absorption capacity is even lower, although yarn quality is the best[7].

3.2.1   Weft yarn preparation for denim fabric

In order to produce good quality denim, the yarn quality used for denim production should be optimal. Thick and thin places in the yarn may lead to prominent defects in the fabric, as denim is a contrast fabric. In the early 1990s, the majority of yarns used in denim production were OE yarns. However, now a days, there is a strong demand for using carded ring spun yarns in both warp and weft, which gives the fabric a softer handle. Warp yarns for bottom weight jeans typically range from 4's Ne to 14's Ne or as per the requirement of finished denim fabric. Finer yarns are used for lighter weight jeans, shirting, and skirts in the range from 16's Ne to 30's Ne. Denim fabric is presently made by a different the combination of ring and open-end yarn (OE) such as ring/ring, OE/OE, and ring/OE," Weaving with combination of ring-spun yarn and open-end yarns can help to reduce fabric costs while still maintaining some favorable ring-spun fabric characteristics. With modern attachments in spinning, few structured denim yarns are also introduced in the market. Using these devices, OE yarns can have a more ring-spun like appearance, and ring-spun yarns can have an increased rough or "antiqued" quality. Yarn spinners can design patterns and effects specific to their needs, which can be downloaded into the machine's electronic control system[7].

3.2.1.1   Modification in ring frame

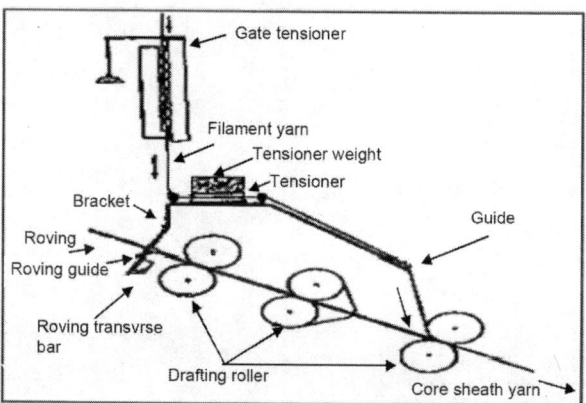

Figure 3.1: Conventional core spinning attachment in ring frame

In conventional core spinning attachment in ring frame machines (Figure 3.1), core-spun yarns can be produced by incorporating a spandex/lycra filament at the back of the front drafting roll of the machine. The lycra core ring spun yarn is produced when the drafted cotton fibers twist around the lycra filament yarn. Elastic core yarns can be produced in ring spinning which consists of an elastic core around which staple fibers are wound (Figure 3.2). The elastic filament is pre-drafted approx 2.5 to 4 fold before being inserted to the spinning zone ahead of the front top rollers. This process has the advantage that the sensitive filament is completely wrapped by staple fibers (Figure 3.3). Lou et al. produced a polyester core-spun yarn containing lycra fibers using a self-designed, multi-section drawing frame and a ring spinning frame. The mechanical properties of the elastic core-spun yarns were examined under various processing conditions. They optimized the draw ratio to enhance the breaking tenacity and elongation of the core-spun elastic yarns.

Figure 3.2: Elastic core spun yarn

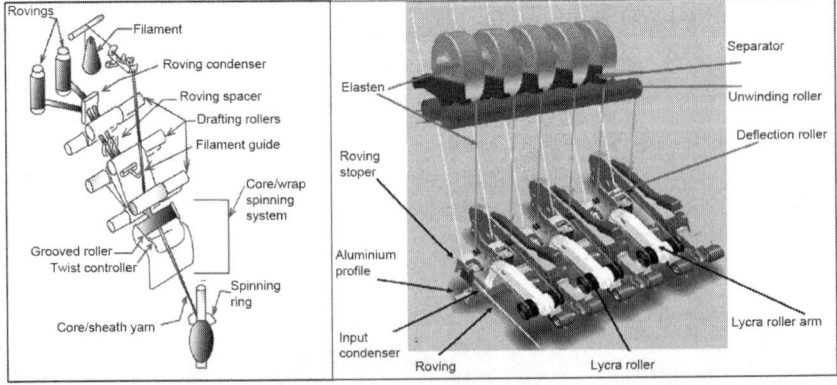

Figure 3.3: Process Layout of manufacturing of core spun yarn in modified ring frame

3.2.2   Warp yarn preparation for denim fabric

At present there are three different types of preparation and dyeing systems for warp yarn denim, which are carried out by the following techniques:

(a) Rope Dyeing

(b) Slasher Dyeing

(c) Loop Dyeing

A schematic process layout of the above three is shown in Figure 3.4. The first system used for denim was rope, while slasher and loop were incorporated later. These types of continuous process systems differ in two aspects:

• The way of preparation and transportation of the yarn inside the equipment.

• The mechanical construction of the equipment.

Before entering the dyeing range the yarn has to be prepared in the form of big cylindrical structures called beams, where the threads are wound in optimal conditions of pressure and regularity. This process, which is begun by spinning cones, is called warping (Figure 3.5). Depending on the type of dyeing equipment, beams are constituted from yarn set parallel, flat, and open in the case of slasher and loop, or arranged in small groups forming ropes, as in the case of rope dyeing (Figure 3.6)[7]. Archroma in a Denim book has mentioned: "Slasher and rope systems represent at least 95% of worldwide denim production".

3.2.2.1   Rope Dyeing

In rope dyeing, ball warps are continuously fed into the rope or chain-dyeing range for application of the indigo dyeing. Typically, 12-36 individual ropes of yarn are fed side-by-side simultaneously into the range. The ropes are kept separate from each other throughout the various parts of the dye range (Figure 3.7). For example, if the total number of ends on the loom beam is3, 456 and each ball would have 288 ends, then the dye set would have a total of 12 ball warps. If there can only be a multiple of 10 balls on the dye range, then there would be 345 ends on 9 balls and 351 ends on the tenth ball. The ropes are first fed into one or more scouring baths, which consist of wetting agents, detergents, and caustic. The purpose of these baths is to remove naturally occurring impurities found on the cotton fiber such as dirt, minerals, ash, pectin, and naturally occurring waxes. It is very important to remove these materials to guarantee uniform wetting and uniform dyeing. The ropes are subsequently fed into one or more water rinsing baths. If a sulfur bottom is required at this point, the ropes of yarn are fed into a bath of a reduced sulfur dye. Similar to indigo, sulfur dyes are water-insoluble. They must be reduced to a water-soluble form before applying to cotton. Unlike indigo, the sulfur dye can penetrate into the core of the cotton fiber/yarn. The purpose of this process is to give the indigo dyed yarns a much deeper and darker shade or to slightly change the shade of the blue yarn to make it unique. Once the reduced sulfur dye is applied to the ropes, they are skied to allow the dye to oxidize into its normal water-insoluble form. The ropes of yarn are then fed into the indigo dye baths and skied after each dip. The ropes of yarn are rinsed in several water baths to remove any unfixed dye.

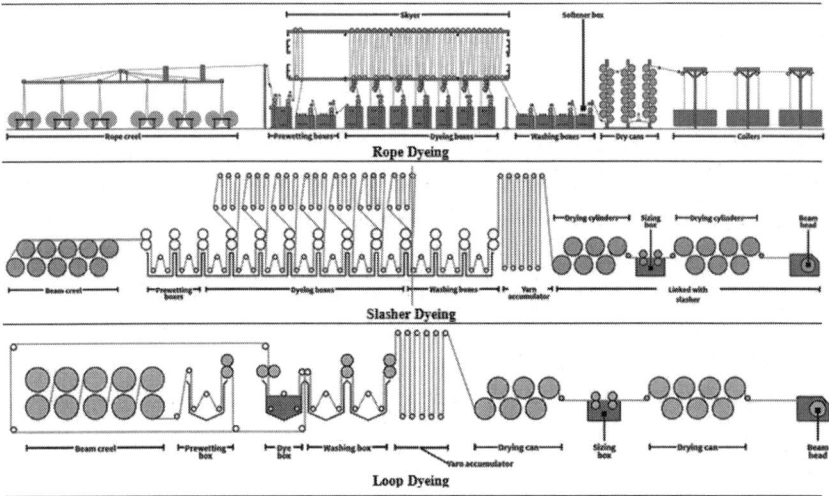

Figure 3.4: Schematic layout warp yarn dyeing for Denim fabric [8]

If a variant type of yarn color is desired, the sulfur dye can be added at this point. Similar to the bottom-dyeing process discussed previously, this process is known as a sulfur top. Although the sulfur dye will migrate towards the core of the fiber/yarn, the sulfur top gives a different type of yarn color performance when garment washed than a sulfur bottom.

The sulfur top process is then followed by a water rinse to remove any unfixed dye. After either rinsing following indigo dyeing or rinsing following sulfur topping, the yarn ropes pass through squeeze rolls to mechanically extract water. The yarns are then dried and coiled into large tubs. The typical type of drying apparatus is a multiple stacks of drying cans. These metal cylinders, which in most cases are Teflon® covered to prevent the yarn from sticking, are filled with steam under pressure. Maintaining a consistent pressure of steam within the cylinder can accurately control the temperature of the surface of each cylinder. Care must be taken not to attempt to dry the rope of yarn too quickly, which causes the dye to migrate to the surface of the rope. Additionally, if the surface of the drying can is too hot, the yarn can be overstressed producing an undesirable glazed appearance that reduces absorbency in later processing. Over-drying of the yarns will weaken them considerably adversely affecting re-beaming, slashing, and weaving. After drying, the color of the yarn is checked either visually or instrumentally. With many modern indigo dye ranges, the color of the yarn is continuously monitored by instruments, which are electronically linked to the controls of the indigo dye baths. This type of control system can automatically adjust the dynamics of the process to obtain the most consistent color from the beginning to the end of the many thousands of yards of yarn contained within a single dye lot. In order to minimize the color variability between denim fabric panels after garment washing, denim manufacturers employ a technique known as sequential dyeing. Basically, this method is based on the concept that the color properties of indigo-dyed yarn processed at a specific time, most closely resemble the color properties of the indigo yarn processed just before and just after that lot. This method has proven much more effective at minimizing color variability in garment washing when compared to the technique of shade sorting alone [8].

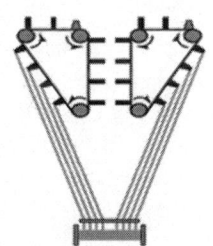

Beam warping (Slasher and loop)

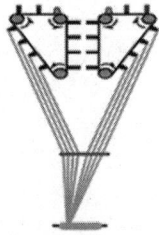

Ball warping (rope)

Figure 3.5: Types of Warp yarn preparation (left)

Figure 3.6: Ball warp in rope form ready to dye (right) [8]

Figure 3.7: View of ropes in rope dyeing [8]

3.2.2.2   Slasher dyeing

For certain manufacturers, the rope or chain dyeing of indigo is not possible or desirable. Many different types of equipment have been tried as an alternative to the dip and sky method of the rope dye range. For some producers, slasher dyeing has become a reasonable alternative method. A slasher is a range normally employed to apply size formulations onto warp yarns before weaving. This range (Figure 3.8), when used for slasher dyeing of indigo, consists of section beams of warp yarn, which are forced into a sheet of yarn. This yarn sheet is then fed into a scouring section where natural impurities are removed. The next section is where indigo dye is applied. In order to achieve fairly deep shades, the indigo is applied in a series of multiple dip and sky applications to allow for shade build up. If the arrangement of the slasher dyeing does not allow for multiple dips and sky applications; then only light and medium shades can be obtained from indigo dyes. The dye application is followed by after washing and drying. With some machinery arrangement, the warp size for weaving is immediately applied. In other arrangements, the warp size is applied onto the yarns employing a separate range. slasher dyeing ranges have a number of advantages and unique characteristics. Slasher dyeing employs a sheet of yarn, which is wound directly onto a warp beam rather than ropes of yarn, which then requires additional handling. This type of dyeing works well with lightweight types of denim. In general, these machines require less floor space, enable smaller production runs, have a quicker turn over time, and are more flexible in their response to changes in the market. As an overall process, these ranges have lower machinery cost; therefore, lower dye costs are realized for specific fabric types. Additionally, the slasher dyeing technique can be used for other dye types for cotton and thus can produce a wide variety of colors other than indigo blue.

3.2.2.3   Loop dyeing

The system of preparation of beams, as well as the way of circulating yarn in the loop range, is the same as in slasher range, which means that parallel threads enter and circulate in one layer. In loop range, as opposed to slasher, yarn does not circulate along with a machine composed of various boxes. After the impregnation with a solution of dyestuff and squeeze, the yarns pass through a closed circuit to enter again into the single vat. This lap of yarn passes as many times as required through the single dyestuff impregnation vat. The number of times the lap passes through the same indigo vat is determined by various factors: the type of yarn used, the number of this yarn, the number of yarn forming the warp, as well as the desired color intensity. It is necessary to maintain precise control during squeezing since during each pass, threads from different layers are being accumulated and all of them have to be squeezed out in an absolutely regular and homogeneous way in the squeezing mangle. Once the warp sheet has passed through the dye vat, the latter turns to the final skying time to oxidize the indigo dyestuff, then washing, drying and sizing, in the form of a single layer (like in the slasher).

3.2.2.4   Analysis of the ranges

In each of the three ranges, it is possible to apply sulfur dyestuffs without restrictions of intensity or fastness (a comparison is given in Table 3.1). However, with the new Indigo colors (under the "Advanced denim" umbrella) alternative to indigo, small ranges begin to appear on the market. They consist of only one dye padder and four boxes, sufficient for sulfur dyeing application and having the advantage of reducing water consumption. They also have more flexibility with smaller and varied batches, which are a rising tendency in the fashion market. In case of both open, one-layer slasher type ranges as well as multilayer ones such as loop there is a possibility of working with dye vats under nitrogen atmosphere. This is done in the effort to maintain homogenous chemical conditions in the bath for a long time after the impregnation of the dye. It improves the quality of dyeing and fastness properties of the dyed yarn. It also requires less consumption of chemical products. Oxidation time (air passage o) which the indigo needs for its complete oxidation (air oxidation), at the speed at which ropes are usually transported, is approximately 4-6 times longer than the duration of the rope bath contact. This period depends also on other factors such as the degree of squeezing of each rope in the exit from the dye vat in the squeezing mangle and the speed of the range. In the case of sulfur dyes, oxidation time is not of great importance because as a general norm our dyes need a chemical (type) oxidation in order to completely develop the color. In rope type range, and especially in case of mercerized yarn, it is common to use a lubricant/softener in the last vat before drying in order to diminish friction between the yarns and to facilitate the opening of the rope (re-beaming). This action facilitates the separation of one thread from another inside the rope. It also supports subsequent operations such as the formation of the beam (the rope is opened and later the ends are rolled out flatly on a cylindrical surface the warp beam) and the chemical finish (sizing).

Table 3.1: Comparison of ranges for Warp dyeing

Range Type	Advantages	Disadvantages
Rope	• High productivity • No side-centre variations • Low waste of thread • No time lost during lot change • Higher intensities of Indigo	• Low flexibility • Taking up a lot of space • The necessity of employing an additional step of opening ropes after dyeing
Slasher	• More compact design • Flexibility in dyeing processes • Continuous process • Possibility of adaptation of the machine in order to obtain more superficial or ring sulfur dyeing	• Risk of selvedge-centre variation • Greater risk of thread rupture • Thread loss during the change of article • Limitation in high intensities of indigo
Loop	• Very compact system • Minimal consumption of chemical products, dyestuffs, and water	• Risk of selvedge-centre variation • Greater risk of yarn breakages • Thread loss during the change of article • Limitation in high intensities of indigo • Limitations in flexibility as regards different processes and dyeing methods

3.2.3   Indigo and Sulphur Dyeing

Indigo is an insoluble pigment, without affinity for cellulose in the oxidized state. For application, the dyestuff should be in reduced leuco alkaline, soluble state. In order to maintain the reduced conditions of indigo, a particular concentration of alkali is used (usually caustic soda) as well as a strong reducing agent, the sodium hydrosulfite, as well as auxiliaries' like dispersing or wetting agents, etc. Indigo dyebath should be controlled by means of chemical parameters such as pH and reduction. Significant changes in these values could make way for variations in the reaction's oxidation-reduction kinetics, which in turn might lead to differences in the diffusion of color, shade, and intensity (Figure 3.8).

The indigo is applied to the yarn by means of repeated impregnations and then passed through the skyer to become gradually oxidized. Contrary to sulfur dyes, indigo is characterized by a low affinity and a quick oxidation tendency. After impregnation and squeezing in the squeezing mangle, the yarn should spend some time in the air duct, where the indigo dye becomes gradually oxidized[9-11]. It becomes insoluble and fixed on the yarn by means of weak bonds (e.g. Van der Waals bonds). This way the indigo is deposited in as many layers on the yarn, as impregnations take place. In any case, the final effect of indigo on the yarn is superficial, due to the low diffusion of the dye (Figure 3.9). This characteristic implies certain limitations as far as fastness is concerned, especially when high-intensity colors are required. In addition, this becomes a resource to get fashion fading of denim with subsequent washing (Figure 3.10).

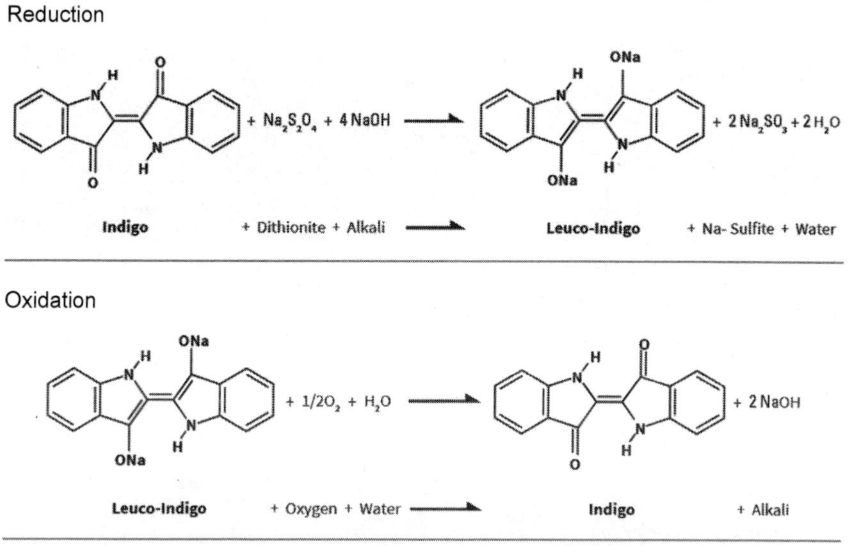

Reduction

Indigo   + Dithionite + Alkali ⟶   Leuco-Indigo   + Na- Sulfite + Water

Oxidation

Leuco-Indigo   + Oxygen + Water ⟶   Indigo   + Alkali

Figure 3.8: Indigo molecule in its reduced form and after oxidation [8]

Figure 3.9: Cross-section of indigo dyed yarns

Figure 3.10: Fading of denim fabric

The sulphur dye molecules are often long (Figure 3.11) and contain few solubilizing groups. In fact, just as in the case of indigo, sulfur dyestuffs are insoluble in their "pigmentary" form, which is their oxidized state. The application of these dyestuffs to cellulose fibers is based on the oxidation-reduction balance of their molecules[12].

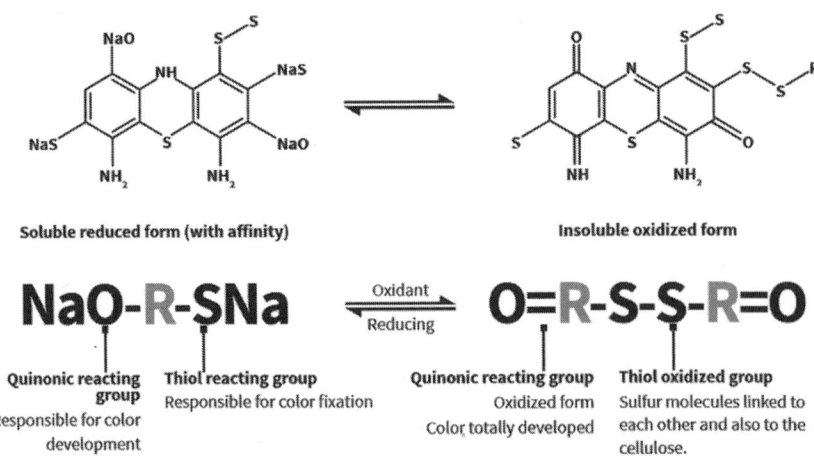

Figure 3.11: Sulfur dye molecule in its reduced and oxidation form [8]

3.3   Variety of denim

The most usual weaving structure used in denim fabric is Twill. Denim is available in different weights ranging from 200-300 g/m^{2}which are categorized as light denim to 300-600 g/m^2, known as heavy denim. Denim is a good choice for casual jackets, skirts, and jeans. Today, most of the jeans we make do not see much hard work, and being comfortable has become a higher priority. These days, most of us prefer a leaner and more figure-hugging cut. So what do you choose- Coolness or comfort? In fact, there is a third option-Indigo Knit fabric, also known as knitted denim i.e. you can have the look of denim with the comfort of sweatpants. This special fabric allows working with the same treatments that we use for our traditional denim fabrics to create the naturally worn-in look. There are treatments such as 3D shaping, breaks and repairs, stone washing as well as different fading agents also available in the market. In recent years, the advancements in garment-finishing techniques have led to easy processing and subsequently, its use has broadened into different lifestyles [13,14]. Several different finishes or washes can be applied to jeans to achieve different looks. Many of the washes aim to give the jeans a worn and torn look. Denim is also available in many colours (Figure 3.12) with more and more inventions in the dyeing of denim such as colours in indigo, sulphur dyes, topping and bottoming with sulphur. Denim is also adored with digital prints and thermal engraving. As a substitute to washing, laser methods are used and have an advantage over conventional washing methods due to their less water and chemical usage. In this section, we have discussed various types of denim fabric finishing like chemical washing techniques, mechanical treatments and printing of denim fabric. Broadly, finishing can be divided into three groups: Denim fabric finishing, Special denim fabric finishing, and denim garment finishing.

3.3.1   Denim fabric finishing

After weaving of denim fabric basic operations like brushing and singeing are carried out. Next vital treatment is for shrinkage control of denim fabric. Shrinkage is required to prevent turning of seams in the direction of the twill structure once the garment has been tailored and washed. Sanforizing is a physical treatment carried out on the denim fabric. It is technically called control of shrinkage due to compression. Both, weft as well as warp yarn undergo a series of movements which allow controlled shrinkage. There are some optional operations as well which are done as per requirement from design, tailoring or customer. These operations are desizing (to remove sizing agent which otherwise is removed during garment enzyme wash), stenter (when denim fabric contains a percentage of elastomeric fiber or other synthetic fibres), mercerization (to generate a shiny look in the denim fabric) and calendaring (to make the structure more compact, smooth and shiny similar to strong ironing).

Figure 3.12: Various colours for denim [8]

3.2.2   Special denim fabric finishes

These operations give the final fabric certain physical properties contributing to the garment's appearance and comfort. Chemical finishes such as softeners, binders, stiffness, water repellent, wicking, coatings, etc. are used to enhance special hand feels properties and functional properties of the denim fabric.

3.2.3   Special garment finishing

These are physical, chemical or even combined finishing carried out on the garment. They are executed in machines called tumbler or rotary drums, similar to domestic washing machines, but with greater capacity and resistance. Garment finishes have become a cornerstone of fashion effects. During the last few years new mechanical (dry) as well as chemical (wet) effects for garments have been developed. By combining both a series of unique and individual looks in each pair of jeans, a vintage denim effect can be obtained[8]. A nice, appealing handle is often the critical criterion for buying a textile. Consequently, in the research works, the influence of softeners on the change of handle is analyzed[16-19].

3.4   Denim garment finishing

Usually, finishing is applied for already sewn garments. Traditionally, denim jean manufacturers have washed their garments with pumice stones to achieve a soft handle as well as a desirable worn look. The main disadvantages of this procedure are the difficulty of removing residual pumice from processed clothing items and the damage to the equipment by the overload of tumbling stones. The pumice stones and particulate material can also clog machine drainage passages and the drains and sewer lines at the machine site [20]. In spite of these disadvantages, pumice stone is still used on its own in some factories. Due to the problems of pumice, alternative methods for stone-washing of denim fabrics have been developed. Cellulases have been used in denim washing for many years and it has been estimated that about 80 % of denim washing is now done in this way. Still, there are different mechanical processing done on denim garments.

3.4.1   Mechanical or physical effects on garment (Dry Denim Processing)

Different abrasion or distressing techniques are used to give a pair of denim jeans a complete fashion worn outlook (Figure 3.13).

Grinding: Normally it is done on pocket edges and bottom hems by running them against an abrasive surface or a stone.

Tagging/clipping: This effect is achieved by using a swift tag machine with a plastic tag which is attached to the fabric. This way it is possible to obtain high contrast effects after washing at the waist, on the edges of the front and rear pockets as well as on the seam on the underside of the jeans.

Damages/breaks: Controlled warp removal to achieve vintage worn-out effects and breaks. Pen grinding tools are used for the process. It is also possible to make damage holes both in the warp as well as the weft. These effects are obtained manually for unique and individual look for the jeans.

Tie effects (such as tie net): Generally used when dyeing garments with irregular effects transferred to the denim garment by means of stonewash type processes. This way it is possible to obtain abrasion effects in certain areas of the garment.

3D effects with resin applications: permanent creases on specific areas of the denim garment (pocket, heel or the back of the knee area).

Patch and repair: manual processes used to obtain a vintage look and unique and individual effects. The effect consists of tearing the fabric in a certain area and then sewing it again manually or using a sewing machine. The obtained effect is new or used vintage.

Laser effects: Laser marking, whiskers/moustaches. These are special effects generated with a laser beam to imitate creases which are formed naturally while wearing jeans. This type of effects can be created in a predetermined way by using a big variety of designs and sketches.

Local tint staining effects / bleached spots: Diluted solutions of oxidizing agents are sprayed on or applied with a sponge after the wet process neutralizing.

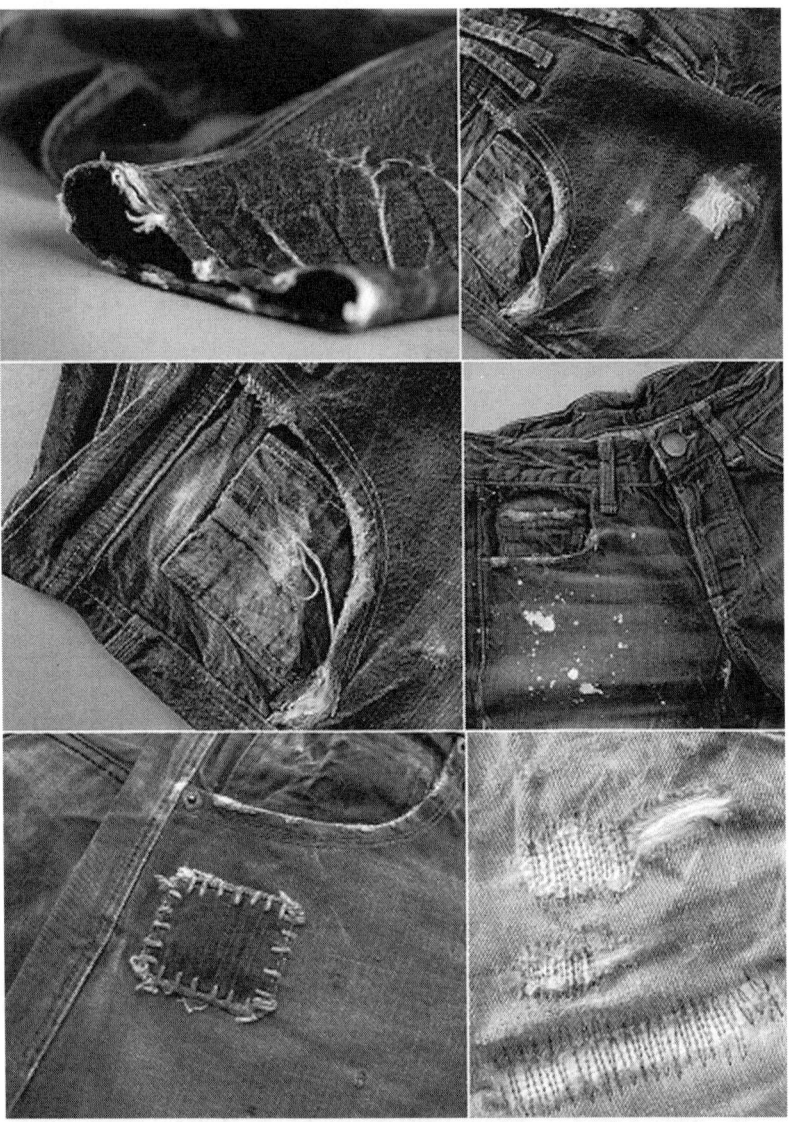

Figure 3.13: Worn out look through mechanical or physical methods

3.4.1   Chemical Effects on Garment (Wet Denim Processing)

Washing is a radical development that has changed the humdrum image of denim from stiff and dull blue fabric to fashion wear. Denim washing nowadays has become a requisite for achieving a final look of any denim garment[1]. Industrial washing involves various new technologies and equipment to obtain desired results. During the whole washing cycle, garments are affected by the entire complex of different factors such as a washing solution, abrasion, creasing, heat, various chemicals, etc. Therefore, intensive destruction of polymers that are the components of fibres takes place and leads to intensive wear of articles. Garments from short fibres feature more intensive wear and tear as in the process of washing fibres are released gradually[15]. Therefore, it is the first important factor to retain the required durability of denim garment with enough processing to achieve the desired wash effect. Secondly, the extensive use of chemicals and water waste has given rise to increased interest in the use of environmentally friendly, nontoxic chemicals for the finishing of denim. In this context, the denim washing industry is striving to develop environmentally friendly washing techniques that can result in zero effluent discharge. Dry treatments or nearly water-free treatments are slowly becoming a sustainable trend for replacing traditional wet treatments in denim garment washing[1].

3.4.2.1   De-sizing

If the denim fabric has not been desized, this process is carried out on the garment with enzymes at the adequate pH according to the type of enzyme used. This procedure can be considered as a wash down technique for denim garments due to the abrasion they suffer in the washing drum.

3.4.2.2   Stone Wash

It is still the most popular of all wash down treatments. Jeans are washed with oval or round pumice stones. The pumice stones are very light and with a rough surface. Stone washing is used on fabrics or garments to produce certain color or texture effects. When the stones come in contact with the fabric, abrasion occurs and superficial coloured fibers are removed. Terms such as "deep stone" or "super stonewash" are an indication of how long the jeans have been stonewashed. The longer the duration of wash, the lighter is the color of the jeans(Figure 3.14).

3.4.2.3   Enzyme Washing

It is a process that uses enzymes to 'stonewash' jeans; it means to imitate the use of stones. The enzyme accelerates the superficial degradation of the fabric, exposing the white core of the color-dyed yarn. It is often used in conjunction with pumice stone to enhance the worn out look. Enzymatic treatment can replace a number of mechanical and chemical operations, which have been applied to improve the comfort and quality of fabrics by now[20,21]. In the textile, industry enzymes are applied mainly to get a cleaner fabric surface with less fuzz, to reduce tendency to pill formation, to improve handle, to smooth the surface combining with traditional softeners. The development studies of this area have been focused on applying enzymes on cellulose materials based on cotton, linen, viscose, and their blends with synthetics fibres[22-27]. Cellulases are enzymes that specifically degrade cellulose, yielding shorter chain cellulose polymers and glucose[28]. During enzymatic treatment removed indigo dye gets redeposit don the white or undyed weft yarn of denim fabric, which is known as back staining process. This process is sometimes required and avoided according to aesthetics. To avoid it, anti-back staining agent can be used here to resist back staining. After washing with enzyme, the trousers are washed off for two times[29]. The continuous research on new enzymes and formulations is going hand in hand with the innovation and sustainability strategies of leading fashion brands and laundries. Laccase enzymes can be used as an alternative to chemical bleaching where the enzymes oxidise indigo to soluble degradation products. Thus, there will be more and more applications for enzymes in denim garment processing, and the further optimisation of existing enzymatic formulations or combining different processes will hold the key for efficient and sustainable washing[1, 30].

3.4.2.4   Bleached effect

Figure 3.14: Stone washed or enzyme treated jeans (left)   **Figure 3.15:** Denim treated with different levels of hydrogen peroxide (Right)

It is another chemical wash down treatment for the denim fabric. A sodium hypochlorite or potassium permanganate as a strong oxidizing agent is normally used when the color is destroyed through the chemical attack. Due to ecological reasons, these kindof treatments will be progressively abandoned in the future. As an alternative, a chemical product like hydrogen peroxide is being used more and more often (Figure 3.15). This treatment can be considered as eco bleach because no hazardous substance remains in the denim fabric once it had been washed and dried.

3.4.2.5   Acid wash effect

A technique of washing jeans achieved by using pumice stones soaked in various chemicals during the stonewashing process. The visual effect is uneven. Double color contrast is obtained after discharged oxidative treatment through combining stable/unstable sulfur dyes with sodium hypochlorite or potassium permanganate.

3.4.2.6   Ozone effect

As a new application technique, this treatment could be the most ecological fashion effect for the denim garment. The color, in the case of sulfur dyes, keeps the initial cast but with a natural fading or bleached effect.

3.4.2.7   Ice blast

Ice Blast is a new way for a denim fabric to rethink the worn look. In this method, the microscopic dry ice crystals are used at a chilling -78°C to blast away the colour at the surface of the yarn. This gives a bright and very natural fade. Technically, the brightness is a result of the colour molecules being immediately removed from the fabric compared to other fading processes such as hand scraping. Since dry ice is made from recycled CO_2 that is turned into ice crystals, the water will evaporate from the fabric, eliminating the need for rinsing the garment in production. That is how we save both water and energy with every pair of Ice Blast jeans[3]. Besides the above finishes, there are many more techniques being used to reform a new look fashion denim garment. In this chapter, we have tried to give a profound vision to gaze the workmanship over a pair of simple denim jeans.

3.5   References

1. Roshan Paul, "Denim: Manufacture, Finishing and Applications" book by Woodhead publishing series in textile : number 164; ISBN 978-0- 85709-843-6.

2. Alexandra Derosa, "What Your Choice in Denim Says About You" Jul 16, 2014 @ 9:15 am

3. Jack & Jones Blog, "Rethinking The Faded Denim Look With Ice Blast" Jeans Intelligence | 18.02.2016

4. Anju Tulshyan and Misba Turk, "DENIM FINISHES" http://www.textilevaluechain.com/index.php/article/industry-general/item/485-denim-finishes

5. Morris M, Prato H (1981) Consumer perception of comfort, fit and tactile characteristics of denim jeans. Textile Chemist and Colorist 13: 24-30.

6. Wu J, Delong M (2006) Chinese perceptions of western-branded denim jeans: a Shanghai case study. Journal of Fashion Marketing and Management: An International Journal 10: 238-250.

7. Nayak R, Padhye R, Dhamija S, Kumar V (2013) Sewability of air-jet textured sewing threads in denim. Journal of Textile and Apparel Technology and Management 8: 1-11.

8. Archroma Life enhanced, Denim Book, from cotton to fashion

9. Xin, J. H., Chong, C. L., and Tu, T., Colour Variation in the Dyeing of Denim Yarn with Indigo, J. Soc. Dyers &Color., 116, 260–265 (2000).

10. Etters, J. N., Indigo Dyeing of Cotton Denim Yarn: Correlating Theory with Practice, J. Soc. Dyers &Color., 109, 251–255 (1993).

11. Chakraborty, J. N., and Chavan, R. B., Dyeing of Denim with Indigo, Indian J. Fibre & Text. Res., 29, 100–109 (2004).

12. M. Gorenšek, MarijaGorjanc and Petra Recelj, Dyeability of Cotton Warp at Dip-Dyeing for Jeans, Textile Research Journal Vol 78(6): 524–531

13. Kumar S, Chatterjee K, Padhye R and Nayak R, Designing and Development of Denim Fabrics: Part 1 - Study the Effect of Fabric Parameters on the Fabric Characteristics for Women's Wear, Journal of Textile Science & Engineering, J Textile SciEng 2016, 6:4

14. Nayak R, Padhye R, Wang L, Chatterjee K, Gupta S (2015) the role of mass customisation in the apparel industry. International Journal of Fashion Design, Technology and Education 8: 162-172.

15. Milda JUCIENĖ, Vaida DOBILAITĖ, Giedrė KAZLAUSKAITĖ, "Influence of Industrial Washing on Denim Properties", ISSN 1392–1320 MATERIALS SCIENCE (MEDŽIAGOTYRA). Vol. 12, No. 4. 2006

16. Buscle-Diller, G., Dong Yang, X. Enzymatic Bleaching of Cotton Fabric with Glucose Oxidase Textile Research Journal 71 (5) 2001: pp. 388 – 394.

17. Webwr, R. New Aspects in Softening. CHT R. Beitlin GMBH 1999: 30 p.

18. Blazevič, P., Strazdienė, E. The Application of Commercial Bleachers for Clothing Decoration Projektowanie, materialy, technologiaskory, odzieży i obuwia Radom, Poland, 2004: pp. 143 – 147.

19. Dobilaitė, V., Jucienė, M. Influence of Industrial Washing on Denim Garment Colours Change Light Industry – Fibrous Materials: III International Scientific Conference Radom, Poland, 2005: pp. 309 – 314.

20. Enzymes. http://www.cht-group.com/

21. Enzymes for Textiles. http://www.mapsenzymes.com/Enzymes_Textile.asp

22. Buschle-Diller, G., Walsh, W. K., Radhakrishnaiah, P., Effect of Enzymatic Treatment on Dyeing and Finishing of Cellulosic Fibers: A Study of the Basic Mechanisms and Optimization of the Process Project: C96-Al National Textile Center Annual Report: November, 1997: pp. 31 – 36.

23. Ciechańska, D., Struszczyk, H., Miettinen-Oinonen, A., Strobin, G. Enzymatic Treatment of Viscose Fibres Based Woven Fabric Fibres & Textiles in Eastern Europe 39 (4) October/December 2002: pp. 60 – 63.

24. Özdil, N., Özdoğan, E., Öktem, T. Effects of Enzymatic Treatment on Various Spun Yarn Fabrics Fibres & Textiles in Eastern Europe 43 (4) 2003: pp. 58 – 61.

25. Pyc, R., Romanowska, I., Galas, E., Sójka-Ledakowicz, J. Hydrolysis of Cellulose Fabrics by Cellulases from Aspergillus IBT-90 Fibres & Textiles in Eastern Europe 24 (1) 1999: pp. 54 – 57.

26. Guzińska, K., Ciechańska, D., Struszczyk, H. Investigation of Biosynthesis Process of Cellulolytic Enzymes for Cellulose Fibre Processing Fibres & Textiles in Eastern Europe 37 (2) 2002: pp. 77 – 81.

27. Onar, N., Saruşik, M. Use of Enzymes and Chitosan Biopolymer in Wool Dyeing Fibres & Textiles in Eastern Europe 49 (1) 2005: pp. 54 – 59.

28. L. Heikinheimo, J. Buchert, A. Miettinen-oinonen and P. Suomine, "Treating Denim Fabrics with TrichodermaReeseiCellulases" Textile Research Journal, Nov 2000, 70: 969

29. M. M. Rahman, "Effects of industrial enzyme wash on denim apparel characteristics", PTJ January 2011, 46-48.

30. Susana Rodríguez Couto* and José Luis Toca-Herrera, "Lacasses in the textile industry", Biotechnology And Molecular Biology Reviews, Article Number - 1C6FD3040214, Vol.1(4), pp. 115-120, December 2006.

4

Analysis of antimicrobial phyto-constituents and finish efficacy of rhizophora mucronata on organic cotton for developing therapeutic apparels

Dr. Krishnaveni Vasudevan*

*Associate Professor, Department of Fashion Technology, Kumaraguru College of Technology, Coimbatore, Tamilnadu, India. Email: krishnassstyle@gmail.com

Abstract : The evolution is taking place due to the simultaneous expansion and improvement of technology in technical textile as well as healthcare division. The patrons and manufacturers are more concern about germ-free lifestyle and there is a necessity of textile products refined with antimicrobial properties. From the natural kingdom, one of the abundant sources of strong natural plant is *Rhizophora mucronata* and it has antiviral, antibacterial, and antifungal properties. The leaf extract was prepared with *Rhizophora mucronata* powder using the solvent ethanol. The plant extract finishing was applied on to the organic cotton fabric by using micro-encapsulation method with natural mordant using Box and Behnken optimized conditions. The phytochemical compounds were identified and screened from the solvent. The anti-bacterial and antifungal property was tested on fabric using AATCC 147 qualitative and AATCC 100 quantitative tests against both grams positive and negative bacterial pathogens with 24 and 48 hour time duration. The antifungal activity was tested on the fabrics against *Aspergillus Niger and Candida albicans* fungi. The test results depict the clear picture and benefits for therapeutic and health care applications.

Keywords : *Rhizophora mucronata*, Organic cotton, Phytochemical, Microorganisms, Antimicrobial finish.

4.1   Introduction

Nature has been a source of miracle plants for thousands of years and a striking number of modern drugs have been isolated from natural resources. Traditional medicines are an important source of potentially useful compounds for the development of chemotherapeutic agents[1]. Medicinal plants play a key role in world health care systems and the serviceable requirements of therapeutic textiles have led to the innovative use of a variety of fibers with enhanced comfort and hygienic properties in the development of new products for medical applications[2]. In recent times, the different ranges of natural fibers and biodegradable polymers are being utilized for developing new products into medical and home textiles. The healthcare textile applications are directly related to the skin and the life of human being, which is required to undergo stringent testing and hygienic criteria. The sustainable plant materials and various parts have been used for developing healthcare products. From that, the *Rhizophora mucronata* plant is one of the miracle sources with wealth and health properties[3,4]. *Rhizophora* plant extracts have been used for centuries as a popular method for treating several health disorders. *Rhizophora mucronata* is biochemically unique, producing a wide array of novel natural products[5]. Plant-derived substances have recently become of great interest owing to their versatile applications.

Rhizophora mucronata plant is one of the Rhizophoraceae family and commonly known as Asiatic mangrove and it is available in the coastal tropical and the subtropical region has been reported to posses several medicinal properties. The leaves, bark, and root of *Rhizophora mucronata* has been used as a traditional medicine in the treatment of diarrhea, dysentery, blood in urine, fever, angina, diabetes, hematuria, and hemorrhage[6,7].

Mangroves that need no introduction in today's world with a variety of bioactive metabolites have been the interest of marine researchers all over the world. Apart from the resources those flourishes in the dense tangle of roots, mud, and tidal water, mangroves are known for its medicinal wealth that has been successfully employed for treating a variety of diseases over hundreds of years[8]. Only the leaves, which have diverse phytochemical compounds, even though, the presence of tannin and saponin are weak positive. The parts of R. mucronata showed some antibacterial activity against S. aureus and E. coli. It is interesting to note that just about all parts showed a broad spectrum of antibacterial activity[9].

Rhizophora mucronata is a famous mangrove plant widely used in the charcoal industry. It has been reported to produce high yields of tannins[10]. In charcoal making, the barks are normally scraped out from the log and left to rot in the field. Mangrove and mangrove associates contain biologically active antiviral, antibacterial, and antifungal compounds[11]. *Rhizophora mucronata* plants are a rich source of Secondary metabolite like steroids, triterpenes, saponins, flavonoids, alkaloids, and, tannins, which play an important role in suppression of deleterious microorganisms[12]. According to the current consumers, more natural with fewer synthetic additive but increase safety and self-life is needed. Resulting from the demands, plants have emerged as popular ingredients and have a tendency of replacing synthetic antimicrobial and antioxidant agents. Hence, the present research work aims at developing anti microbial finished healthcare products by using *Rhizophora mucronata* extract for therapeutic applications.

4.2   Materials and Methods

4.2.1   Materials

4.2.1.1   Selection of material

The plain weave organic cotton fabric with a count of 2 × 30's was chosen for the study. The fabric was processed by different treatments namely desized, scoured, and bleached prior to the application of finish[13].

4.2.1.2   Selection of plant

Rhizophora mucronata is a medicinal mangrove plant used to treat pain, inflammation and reduce blood glucose level in the Southeast coast of India. *R. mucronata* is commonly found in most mangrove swamps in tropical Asia, from the delta of the Indus in Pakistan to Vietnam and Hainan. The leaf and bark extracts of Rhizophora shows antiviral activity against Newcastle diseases and are also used for skin disorders, boils, and wounds. Mangrove plants are a rich source of tannins which take part in the growth control of deleterious microorganisms. The Rhizophora were purchased from Agriculture University, Chidambaram.

4. 2.1.3   Preparation of Rhizophora Mucronata extract

Figure 4.1: Preparation of Rhizophora leaf extract

The Rhizophora leaves were collected, shadow dried and converted into powder form by using automatic equipment. This Rhizophora powder was converted into solution form using the solvents by soxhlet apparatus with the following standard procedure. The 100 gms of herb powder was mixed with 200 ml of ethanol for seven hours using soxhlet apparatus by hot extraction method for extraction of solution. After the extraction process, the solution was kept for solvent evaporation for about 8 hours. The residual precipitate of the extract was stored in a refrigerator with the use of air-tight containers at 5°C. Based on the necessitate of requirement, the precipitate of the extract was diluted and utilized for further end use[14,15].

4.2.1.4   Method of antimicrobial finishing of extract on organic cotton fabric

The organic cotton samples were finishing with 5% and 15 % optimized concentration of herbal extract using the material liquor ratio of 1:10 with optimized conditions namely 20 pascal pressure at 50°C temperature for about 1 hr time duration with pomegranate mordant as a cross-linking agent. The extracts were applied to the organic cotton fabric by microencapsulation technique. Microencapsulation was done using Rhizophora as core material and gum acacia as wall material. Fourteen gram of wall material was allowed to swell for half an hour by mixing with 100 ml hot water. To this mixture, 50 ml of hot water was added and stirred for 15 min maintaining the temperature between 45°C. 140 ml of the core material was added and stirred at 400 rpm for further 20 mins followed by dropwise addition of sodium sulphate for 5 mins. Then the stirrer speed was reduced for a particular time period than the 17% of formaldehyde was added to the concrete solution. After the preparation of solution mixture, the organic cotton fabric was immersed in the solution for about 2 hours and then it was applied on fabric using padding mangle and dried at 85°C in the oven for 10 mins and then cured at 140 °C for 2 minutes.

4.2.2   Methods

The different methods are used to identify the phytochemical constituents present in the *Rhizophora mucronata* herbal extract which represents the antimicrobial activity on the solution and extract coated fabric samples.

4.2.2.1   Method of identifying Phyto -constitutents in plant extract

The phytochemical constituents are commonly identified by various qualitative methods used for each active phytoconstituents present in the herbal mucronata extract namely alkaloids, flavanoids, phenols, and tannins compounds. The following standard test methods were used for identification of each phyto-compounds.

(a) Wagner's test method for Alkaloids

The 1.5% v/v of hydrochloric acid was taken in a beaker and added a few drops of Wagner's reagent to acidify the 1.5 ml of herbal extract for one hour time. After the time period, the formed precipitate was noted and then reddish precipitate which represents the presence of alkaloids compounds in extract solution.

(b) Alkaline Reagent test method for Flavanoids

A little amount of herbal extract was taken in a conical flask and five drops of sodium hydroxide solution were added to this mixture and waited for a few seconds. Noted the colour changes in the solution and the Intense yellow color was formed. Then add a few drops of dilute hydrochloric acid in the colored solution and a few seconds the colored solution was turned into a colorless, which indicates the presence of flavanoid compounds.

(c) Lead Acetate test method for Phenols

The Rhizophora mucronata extract of 1.0 ml was diluted with 2.0 ml of distilled water. Then add a few drops of 1% aqueous lead acetate solution to that mixture. Then the subsequent time period the changes in the solution was noted. The yellow precipitate was formed and it designates the occurrence of phenolic compounds in the extract.

(d) Ferric Chloride test method for Tannins

A small amount of aqueous 5% Ferric chloride was mixed with 1.0 ml of the herbal extract in a beaker and the changes were noted. After a few minutes, the solution was turned in to bluish black color then a few seconds the color gets disappeared. After that, pour a few drops of dilute sulphuric acid in the mixture and a yellowish brown precipitate was formed which be a sign of the presence of tannins[16-18].

4.2.2.2   *Assessment method of antibacterial activity on Rhizophora mucronata ethanolic extract finished fabric (AATCC-Test Method)*

The antibacterial activity of the *Rhizophora mucronata* leaf extract was analyzed by using both the AATCC-147-Qualitative Agar well diffusion test method and Quantitative method of Broth dilution test (AATCC-100).

Preparation of gram positive and negative bacterial cultures

The four different bacterial cultures and two fungal cultures were developed from Microbial Type Culture Collection (MTCC), Department of Biotechnology Laboratory, Kumaraguru College of Technology, Coimbatore, Tamil Nadu, India. The developed bacterial cultures for the study were namely gram-positive bacterial pathogens *Staphylococcus aureus* (MTCC–724) and *Klebsiella pneumoniae (*MTCC-107), gram-negative bacterial pathogens namely *Pseudomonas auruginosa* (MTCC-421) and *Escherichia coli*(MTCC-442) respectively. The developed bacterial cultures were maintained on nutrient agar slant and stored separately in a refrigerator at 4°C.

(a) Qualitative antibacterial activity assessment by Agar well diffusion method (AATCC- 147)

The efficacy of antibacterial against microbial growth was analyzed by using agar well diffusion standard test method on *Rhizophora mucronata* leaf extract coated fabrics[22]. A 25 ml of nutrient agar was prepared and permitted for sterilization at 121°C for about 20 minutes. Then the Petri plates were autoclaved in a hot air oven at 121°C for 45 minutes. After that, the ethanolic solution has been converted into 120 μg/ml concentration for the use. The 25 ml of nutrient agar was poured into the Petri plates and was allowed to solidify for a particular standard time period. Then the herbal extract was added in the developed well and the plates were incubated for 24 hours at 37°C. After 24 hours, the antibacterial activity was assessed against the gram positive and gram-negative organisms namely *Staphylococcus aureus*, *Klebsiella pneumoniae, Pseudomonas auruginosa,* and *Escherichia coli* by screening the zone of inhibition[19].

(b) Quantitative antibacterial activity assessment by broth dilution test method (AATCC -100)

A 25 ml of Nutrient broth powder was weighed and poured into a conical flask and add a required amount of distilled water and stirred it well. Then the 150 μl Staphylococcus aureus bacteria were added into two conical flasks, which contain the *Rhizophora mucronata* leaf extract and standard culture. Similarly, 150 μl Escherichia coli bacteria were also added into two conical flasks as per the standard norms. After that, the flasks were kept in a shaker for 24 hours under average speed at room temperature. Then the 3 gms of Nutrient broth was taken and diluted with distilled water for zero calibration. Before calculating the readings, the machine was calibrated at 600 nm and the readings were noted and then reduction bacterial percentage of the absorption values was calculated against standard penicillin antibiotic.

4.2.2.3   Assessment method of antifungal activity on Rhizophora mucronata ethanolic extract finished fabrics (AATCC-Test Method)

Antifungal activity of the *Rhizophora mucronata* leaf extract was analyzed by using the AATCC-147-Qualitative Agar well diffusion test method against the standard fungi.

(a) Antifungal activity assessment by agar well diffusion method

The potato dextrose agar was prepared and it's allowed for sterilization at 121°C for about 20 minutes. Then the selected Petri plates were autoclaved in a hot air oven at 121°C for 45 minutes and the ethanolic extract was converted into 100 µg/ml concentration for the required purpose. The 20 ml of potato dextrose agar was poured into the Petri plates and were allowed to solidify for a standard time period. Then the plant extract was added in the developed well and the plates were incubated for 62 hours at 36°C [19]. After 24 hours, the antifungal activity was analyzed by measuring the zone of inhibition against the test organisms such as *Aspergillus brasiliensis* and *Aspergillus fumigates*. [20-22]

4.3   Results and discussion

The phytochemical constituents screening and antimicrobial activity efficacy test results were discussed in this headings.

4.3.1   Identification of Phyto- constituents in herbal extract

The standard qualitative phyto chemical screening of *Rhizophora mucronata* leaf extract test results are shown in Table 4.1.

Table 4.1: Qualitative Phytochemical analysis of Rhizophora mucronata extract

S. No.	Plant constituents	Leaf extract
1	Alkaloids	
	(a) Wagner's test	+ (Strong)
2	Flavanoids	
	(a) Alkaline reagent test	+ (Strong)
3	Phenol	
	(a) Lead acetate test	+ (Strong)
4	Tannins	
	(a) Ferric chloride test	+- (weak)

The test results proved the occurrence of phytochemical constituents in the *Rhizophora mucronata* leaf extract namely alkaloids, flavanoids, and phenols in a strong manner whereas tannins were in a weak spot but it does not affect the efficacy. In all the test, the precipitate color represents the presence of these above-mentioned components induces either individually or in combination to posse's antimicrobial activity. Flavonoids are found to be active antimicrobial component against a wide range of microorganisms.

3.2 Standard antimicrobial activity Assessment method of Rhizophora mucronata leaf extract finished fabrics (AATCC-147 and AATCC 100 Test Method)

The qualitative and quantitative antibacterial activity of the *Rhizophora* mucronata leaf extract has been shown in the Table 2 and 3.

(a) Qualitative antibacterial activity assessment by agar well diffusion method

The zone of inhibition results of *Rhizophora* mucronata leaf extract against gram-positive bacterial pathogens namely *Klebsiella pneumoniae, Staphylococcus aureus,* and gram-negative bacterial pathogens namely *Pseudomonas auruginosa* and *Escherichia coli* by agar well diffusion method were shown in Table 4.2 and Figure 4.2.

Table 4.2: Antibacterial zone of inhibition in (mm) against gram positive and gram-negative bacterial pathogens on ethanolic leaf extract of Rhizophora mucronata

S. No	Samples	Antibacterial activity (Zone of inhibition in mm) against *Staphylococcus aureus*-MTCC 724	Antibacterial activity (Zone of inhibition in mm) against *Klebsiella pneumoniae*-MTCC 107	Antibacterial activity (Zone of inhibition in mm) against *Pseudomonas auruginosa* - MTCC 421	Antibacterial activity (Zone of inhibition in mm) against *Escherichia coli*-MTCC 442
1	5% conc *Rhizophora* extract coated sample	22	19	16	17
2	15% conc *Rhizophora* extract coated sample	28	24	20	18

Table 4.2: Assessment of antibacterial activity zone of inhibition in mm

From the table 4.2, it has been found that the zone of inhibition results of leaf extract showed good antibacterial activity against gram-positive pathogens namely *Staphylococcus aureus* (28 mm) and *Klebsiella pneumoniae* (24 mm) than gram-negative pathogens namely *Pseudomonas auruginosa* (20 mm) and *Escherichia coli* (18 mm). The extract proved that it has very good bacterial growth control over the positive pathogens than compared to the negative pathogens.

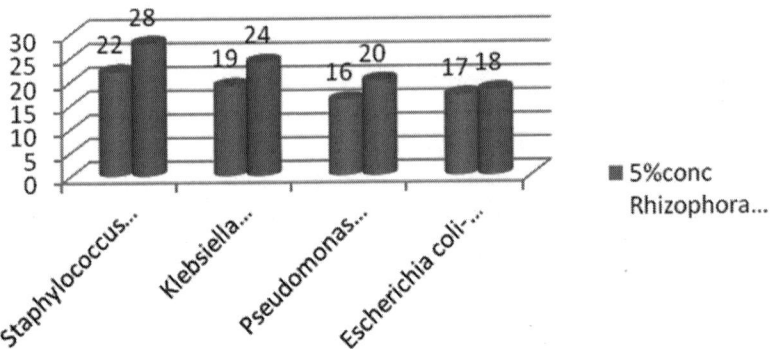

Figure 4.2: Antibacterial efficacy of 5% conc Rhizophora extract coated sample

(b) Quantitative antibacterial activity assessment by broth dilution method

The *Rhizophora mucronata* leaf extracts bacterial growth by broth dilution quantitative method was shown in table 4.3. The absorbance value was measured at 600 nm on the conical flasks for both gram-positive and negative pathogens namely *Staphylococcus aureus* and *Escheirchia Coli.*

Table 4.3: Quantitative analysis of test results of Rhizophora mucronata extract (Broth dilution test)

Samples	Antibacterial activity (Absorbance value OD at 600 nm)			
	Staphylococcus aureus gram positive bacteria (nm)		*Escheirchia Coli* gram negative bacteria (nm)	
	24 hrs	**48 hrs**	**24 hrs**	**48 hrs**
Rhizophora mucronata leaf extract standard concentration	1.61	1.80	1.46	1.70
Standard penicillin Antibiotic	1.50	1.71	1.36	1.53

From the Table 4.3, the quantitative result shows that the *Rhizophora mucronata* leaf extract has good absorbance value in 24 hrs and 48 hrs time treatment against gram negative and positive bacteria when compared to the standard antibiotic. The OD value of *Rhizophora mucronata* extract showed that better bacterial reduction percentage and good bacterial activity against gram negative E.Coli (1.46 nm & 1.70 nm) than gram-positive Staphylococcus aureus (1.61 nm & 1.80 nm) bacteria strains.

4.3.3 Qualitative anti fungal activity assessment by agar well diffusion method

The antifungal activity of Rhizophora has been performed against the microorganisms namely *Aspergillus niger and Candida albicans*. The test result has been shown in Table 4.4.

Table 4.4: Antifungal activity of Rhizophora extract

S. No	Samples	Antifungal activity (Zone of inhibition in mm) against Aspergillus niger	Antifungal activity (Zone of inhibition in mm) against Candida albicans
1	5% conc *Rhizophora* extract coated sample	14	12
2	15% conc *Rhizophora* extract coated sample	20	16
3	Standard Antibiotic chloramphenicol	22	17

From Table 4.4, it has been pointed out that the *Rhizophora* leaf extract of 15% concentration of coated fabric shows the highest antifungal activity when compared to 5% concentration against *Aspergillus niger* than *Candida albicans*. The *Rhizophora* leaf extract showed good control over the fungi growth similar to standard chloramphenicol Antibiotic.

4.4 Conclusion

The conclusion of test results, the Rhizophora extract showed an excellent effect of antimicrobial property against pathogenic bacteria and fungi. The phytochemical constituents present in the herbal extract such as flavanoids, alkaloids, phenol, and tannins. The presence of both qualitative and quantitative antimicrobial test results also showed greater levels of antibacterial activity against gram-positive bacterial pathogens namely *Staphylococcus aureus* and *Klebsiella pneumoniae* than gram-negative bacterial pathogens *Pseudomonas auruginosa* and *Escherichia coli*. The OD value of extract showed that better bacterial reduction percentage and good activity against gram-negative E.Coli than gram-positive Staphylococcus aureus bacterial strains. Based on the phytochemical screening and antimicrobial efficacy assessments, the *Rhizophora mucronata* leaf extract proved that it will be more suitable for therapeutic applications. Hence, this investigate work will give a nutshell about the sustainable, antimicrobial medicated product developments for the healthcare field as well as raw material is from 100% natural renewable resources and it has foremost sustainable benefits.

4.5   References

1. Devprakash, K.K.Srinivasan, T.Subburaju, Sachin kumar singh. Antimicrobial activity of alcoholic and aqueous extracts of Vetiveria zizanioides. J. Phar Res, 2011; 4(5): 1343-1344.

2. Krishnaveni,V. Investigation of phytochemical and antibacterial activity on agave Americana methanolic extract for medical applications, Int J Phar Bio, 2017; 8(3): (B): 500 –505

3. Rao, R.R and M.R. Suseela. Vetiveria zizaniodes (Linn.) Nash. A multipurpose ecofriendly grass of India. Proc.Second Int. Conf on Vetiver, Office of the Royal Development Projects Board, Bangkok. 2000: 444–448.

4. Sangeetha D., Stella.D.Screening of Antimicrobial Activity of Vetiver Extracts against Certain Pathogenic Microorganisms. Int. J. Pharm Bio Arc. 2012; 3(1):197-203.

5. Chandrasekaran, M., Kannathasan, K., Venkatesalu V., and Prabhakar, K. 2009. Antibacterial activity of some salt marsh halophytes and mangrove plants against methicillin resistant Staphylococcus aureus. Wo J.lMicr Biot. 2014; 25:155-160

6. Kusuma.S, Anil Kumar.P, Boopalan.K, Potent antimicrobial activity of Rhizophora mucronata, J. Ecobio, 2011; 3(11): 40-41.

7. Duke, J.A. Wain, K.K, Medicinal plants of the world. Computer index with more than 85,000 entries- Mangroves: The Forgotten Forest between Land and Sea. Tropical Press, 1981; [3]: 97.

8. Krishnaveni, V, Srinivasan, J, Investigation of phytochemical and anti-bacterial activity on Rhizophora apiculata ethanolic extract for medical textile applications," A. J. Micr, Bio Env Sci,2017; 19(11): S8-S11.

9. Nazima.B, Noshin.I, Asiatic Mangrove (Rhizophora mucronata) – An overview, Eu Ace Res, 2014;2,(3) : 75-81.

10. Jain, K., Afidah, A. R. & Mohd Azman, I, Anti-corrosive performance of wash primer based on mangrove tannin. in Proceedings of the 15th Symposium of Malaysian Chemical Engineering. Universiti Teknologi Malaysia, Skudai. 2002: 323–327

11. Lim. S. H, Darah. I, Jain. K, Antimicrobial activities of tannins extracted from Rhizaphora apiculata barks ,J. Trop Fo Sci ,2006; 18(1): 59--65 .

12. Tasleem Arif ,J.D,Bhosale,NareshKumar,T.K, Mandal,R.S, Bendre,G.S, Lavekar, K, Natural products – antifungal agents derived from plants, J. A Nat Pro Res, 2009;1(7):230-243.

13. Krishnaveni, V; Ampritha, S.,Analysis of Phytochemical And Anti-Microbial Activity On Vetiveriazizanioides Ethanolic Extract For Healthcare Applications" 2016,I J Phar Bio Sci, 7(2) (B) :758 – 765.

14. Jayaraman S, Manoharan M S and Illanchezian S. In vitro antimicrobial and anti tumor activities of Stevia rebaudiana. Tr. J Phar Res, 2008. 7(4), 1143-1149.

15. Krishnaveni, V.,Ampritha.S, Application Of Vetiver Extract On Cotton Fabric For Developing Healthcare Products ", in Chapter 3 in the book *on* Sustainability in Fashion and Apparels: Challenges , 2018 ,WPI Publishing , CRC Press, Tayler And Francis Group,: 21-27

16. Ratha, M, Subha. K, Senthilkumar. G and Panneerselvam.A. Screening of phytochemical and antibacterial activity of Hemidesmus indicus (L.) and Vetiveria zizanoides (L.). European J. Experimental Biology.2012; 2 (2):363-368.

17. Snigdha Mishra, Satish Kumar Sharma, Aleza Rizvi, Abhineet Chowdhary. Physico-Chemical Standardization and Phytochemical Screening of Potential Medicinal Herb: Vetiveria Zizanioides (Roots). Int. J. Phytotherapy, 2014; Vol 4 Issue 1: 1-6.

18. Solomon Charles .U, Arukwe Uche. I. Onuoha Ifeanyi. Preliminary phytochemical screening of different solvent extracts of stem bark and roots of Dennetia tripetala G. Baker. Asian J. Plant Sci and Res. 2013; 3(3):10-13.

19. Krishnaveni,V, Investigation Of Phytochemical And Anti-Bacterial Activity On *Agave Americana* Methanolic Extract For Medical Applications," Int J Phar Bio Sci, 2017,8,(3) (B) :500 – 505.

20. Shelar, P. S, Reddy, G, Shelar and Reddy, V. Medicinal value of mangroves and its antimicrobial properties - A review Continental. J. Fish Aqua. Sci, 2012. 6: 26-37

21. Dikshit, A. and Husain, A. 1984. Antifungal action of some essential oils against animal pathogens. Fitoterapia 55: 171-176

22. Hassan A, Rahman S, Deeba F and Mahmud S. Antimicrobial activity of some plant extracts having hepatoprotective effects. J medicinal plant research, 2009. 3 (1), 020-023.

5

Introspection into the eco-friendly garment finishes and its applications

Dr. M. Rameshkumar* Mr. S. Paramasivam* S & Dr. M. Parthiban**

Department of Fashion Technology, Sona College of Technology, Salem – 636 005

E.mail: ramesh@sonatech.ac.in

**Department of Fashion Technology, PSG College of Technology, Coimbatore-641 004*

Email: parthi111180@gmail.com

Abstract : Textile dyeing industries consume a vast amount of chemicals and water and it releases a huge amount of harmful volatile agents that affect the environment. A new concept of fabric finishing with plant extracts having medicinal values is emerging with a fair degree of acceptability. In this paper, an attempt has been made to explore the concepts of eco-friendly textile finishing techniques which includes nano-finishing, enzymatic finishing, plasma treatment, microencapsulation, wrinkle-free finish, flame retardant finish, anti-microbial finish, UV- protection finish, bio-polishing, sandblasting finish, hydrophilic finish, water/oil repellent finish, soil release finish, anti-microbial finishing using bioactive agents like chitosan, sericin, neem extraction and natural dyes. The purpose of the above finishing techniques is to improve the appearance of fabric, feel of the fabric, and drape characteristics without any harm to the user and the environment.

Keywords : Finishing, eco-friendly finish, enzymatic finish, plasma treatment, wrinkle-free finish, washing of garments.

5.1   Introduction

It is well known that every product, which is used by the consumer, has less or more impact on the environment, which the consumer doesn't know. Any product, which is made, used or disposed of in such a way that the product will not create any harm to the environment, could be considered as an eco-friendly product.

In the textile industry, both natural and manufacturedfibers are of equal importance in the application area. Natural fibers such as cotton, silk, wool, linen, jute, hemp, etc., and man-made fibers such as polyester, acrylic, nylon, etc., are the most commonly used fibers for various end uses. A large amount of water and chemicals are consumed in the process of dyeing, printing, and finishing. Along with this, numerous volatile agents are released into the atmosphere that is particularly harmful to human health. This can be stated in a way that the textile processing industry is a prominent factor of water and air pollution.[1]

Textile chemical processing industry uses huge quantities of chemicals, auxiliaries, and dyes. The quality of chemicals towards the environment is not tested due to the high cost involved in the process of conventional animal testing and hence most of the chemicals have been poorly tested for their toxicity. Only bio tests can clearly identify the harmfulness of the chemicals to the living organisms.

Many new advanced technologies are involved in the eco-friendly finishing of textiles. Among these techniques, finishing off by natural products is an interesting technique which gains popularity because of the medicinal value achieved by the finishing of textiles using herb and plant extracts. Different types of properties like anti-itching, anti-microbial, anti-allergic, anti-septic, fragrance, wound healing, freshness, and coolness, etc., can be incorporated into the fabric with the aid of herbal extracts.

In this article, an attempt is made to explore the outcomes of herbal finishing. Basil leaves, pomegranate rind, Aloe Vera, neem, jatropha curcas, etc., are some of the eco-friendly material used in the textile applications.

Many trails have been made in the development of herbal extract from the plants and to produce herbal finished textiles to cure selected disease. Herbs have the potential of curing diseases like psoriasis, allergic dermatitis, asthma, liver disorders, joint pains, headache etc[2]. The medicinal properties of herb can be incorporated into a textile fabric by dyeing and finishing, micro-encapsulation, etc. Most of the natural dyes obtained from plants exhibit strong antimicrobial properties. Therefore, dyeing of fabrics with natural dyes and bioactive plant extract on to cotton fabrics is an emerging technology in the production of medical textiles[3].

Antimicrobial finish not only protects the textile fabric, but also the user from microbial infestation. Hygiene has become the priority on textile fabrics as they are termed as the 'second skin' and are closer to the human body. This aspect calls for greater importance given to anti-microbial finish in textiles[4].

The term Herbal Textile is used for a textile material, which is dyed or finished entirely with herbal extracts without using any sort of chemicals. These herbs are different from vegetable dyes as they are natural and have medicinal value. These herbs have direct application on the fabric with the help of natural ingredients, so that the medicinal value of the herbs can be kept intact. The concept of herbal textiles is derived from Ayurvastra – a branch of Ayurveda, the ancient 5,000-year-old Indian system of Vedic health care. "Ayur" means health, "Veda" means wisdom, and "Vastra" is a cloth or clothing. Ayurvastra clothing, which is made from organic cotton fabric that has been finished with special herbs and oils, promotes health care and cure special diseases depending upon the blends of embedded herbs and oils[5].

5.2   Textile finishing treatment

5.2.1   Plasma treatment of textiles

Textile has now become a domain for interdisciplinary approaches such as the application of nanotechnology, development of smart and conductive fabrics. The techniques used in surface modification, are performed either with chemicals or high energy radiations. Surface modification of textiles using gamma rays and plasma is catching up the research front at a fast pace.

Plasma is defined as a whole or partially ionized gas with an equal number of positively and negatively charged particles. It is often called the "fourth state of matters." Plasma can be created artificially by exposing the gases such as oxygen. This treatment has been employed as an eco-friendly technique to improve the efficacy of textile chemical processing. It is the most attractive as being a clean, simple, and multi-functional process[6].

Plasma exists in two types as low and high temperature plasma. Low temperature plasma is the ionized gases generated at pressures between 0.1 and 2 Torr used in surface modification and organic cleaning. High temperature plasma is the plasma found at atmospheric pressure in its man-made form as the plasma torch in stainless steel deposition or occurring naturally as lightning[7].

5.2.1.1   Advantages of plasma technique

- It is an eco-friendly technique as the high temperature plasma is found at atmospheric pressure in its man-made form as the plasma torch in stainless steel deposition or occurring naturally as lightning assumes low energy and chemicals and there is no problem with the disposal of waste.
- Optimization of the surface properties without altering the bulk characteristics.
- The time required for the treatment is short.
- No chemical products and gases are produced and it is considered as operator friendly technique.
- Applicable to all substrates suitable for vacuum processes.

- The process is performed in a dry, closed system and excels in high reliability and safety.
- The surface properties of the polymers which are unable to modify with wet chemicals can also be changed by using plasma technique[8].

5.2.1.2   Application of plasma in textiles

The plasma modified polymeric materials can be used as textiles, membranes, foils, non-wovens, composites and so on.

5.2.2   Laser treatment

The next physical surface treatment method which creates hydrophilic groups on hydrophobic fibers and enhances the dyeing process is laser treatment. Extensive research has been carried out into the possibility of surface finishing of synthetic fiber fabrics by laser irradiation. A laser type must be selected which irradiates in a strongly absorbing spectral region of the high polymers. It is possible to obtain surface structuring without affecting the thermal and mechanical properties of the body of the fiber. Surface properties that change in the fabric include particle adhesion, wettability, and optical properties[9].

5.2.3   The green alternative

With biopreparation using the enzyme the cotton fibers can be treated under very mild condition. The environmental impact is reduced since there is less chemical waste and a lower volume of water is needed for the procedure. The biopreparation process decreases both effluent load and water usage to the extent that the new technology becomes an economically viable alternative. Enzymes replace the hot sodium hydroxide to remove the impurities and leaving the cotton fiber intact. It is believed that the replacement of caustic scouring of cotton substrates by biopreparation with selected enzymes will result in the following quantifiable improvements: Reduction in BOD, COD, TDS, alkalinity, process time, cotton weight loss, and harshness of hand. An extremely powerful alkaline pectinase recently has been isolated. This new enzyme is now being produced in volume and is being reduced to commercial use in biopreparation on a worldwide basis. The major benefit of this enzyme in biopreparation is that the enzyme does not destroy the cellulose of the cotton fiber. The enzyme is a pectate lyase, and as such very rapidly catalyses the hydrolysis of salts of poly galacturonic acids (pectins) in the primary wall matrix. The term alkaline pectinase is used to describe the enzyme because the biological catalyst is used under mild alkaline conditions, which are very beneficial in the preparation process[10].

5.2.3   Super critical carbon dioxide

Hydrophobic textile materials require creating pores so that the non-ionic dye particles would enter into the textile materials at high temperature and pressure during the dyeing process. After dyeing when the temperature of the dyed materials goes down to the room temperature, the dye particles would be entrapped by the dyed textile materials. Therefore, the hydrophobic textiles are normally dyed from aqueous dye liquors. In such dyeing, complete bath exhaustion never occurs, i.e. the dye does not exhaust quantitatively onto the respective substrate, with the further result that, after the dyeing process, the residual dye liquor still contains more or less amount of dye depending on the particular dyes and substrates. For this reason, dyeing results in the formation of relatively large amount of colored effluents, which have to be purified at considerable trouble and expense.

- The process of the invention has a number of advantages as they claimed such as:
- The supercritical carbon dioxide used in the process does not pass into the effluent, but is reused after the dyeing process. Therefore, no contamination of the effluent occurs.
- Further, compared with the aqueous system, the mass transfer reactions necessary for dyeing the textile substrate proceed substantially faster, so that the textile substrate to be dyed can be penetrated particularly well and rapidly by the dye liquor.
- When dyeing carried out in wound packages by the process of the invention, no unevenness occurs with respect to the penetration of the packages. This unevenness is regarded as responsible for a lot of defects in the conventional process for the beam dyeing of flat goods.
- In addition, the novel process does not give rise to the undesirable agglomeration of disperse dyes which from time to time occurs in conventional dyeing with disperse dyes. Thus, the spotting, which may occur in the conventional dyeing process carried out in aqueous systems, is avoided by using the process of the invention[11].

5.2.4   Ultrasonic assisted wet processing

Ultrasonic represents a special branch of general acoustics, the science of mechanical oscillations of solids, liquids, and gaseous media. With reference to the properties of the human ear, high-frequency inaudible oscillations are ultrasonic or supersonic. In other words, while the normal range of human hearing is in between 16Hz & 16 kHz. Ultrasonic frequencies lie between

20 kHz and 500 MHz Expressed in physical terms, the sound produced by mechanical oscillations of elastic media. The occurrence of sound presupposes the existence of material it can present itself in solid, liquid, or gaseous media. Wet processing of textiles uses large quantities of water, and electrical and thermal energy. Most of these processes involve the use of chemicals for assisting, accelerating, or retarding the reaction rates and carried out at elevated temperatures to transfer mass from processing liquid medium to across the surface of the textile material in a reasonable time. Scaling up from lab scale trials to pilot plant trials is often a challenging process. In order to provide better results during dyeing, high intensities are required. It is very difficult to produce high intensity, uniform ultrasound in a large vessel.

Ultrasound reduces processing time and energy consumption, maintain or improve product quality, and reduce the use of auxiliary chemicals. In essence, the use of ultrasound for dyeing will use electricity to replace expensive thermal energy and chemicals, which have to be treated in wastewater[12].

5.2.5   Bubbling phenomenon

Ultrasound energy is sound waves with frequencies above 20,000 oscillations per second, which is above the upper limit of human hearing. In liquid, these high-frequency waves cause the formation of microscopic bubbles or cavitations. They also cause insignificant heating of the liquid." Ultrasound causes cavitation bubbles to form in the liquid. When the bubbles collapse, they generate tiny but powerful shock waves. We needed to agitate the border layer of liquid to get the liquor through the barrier more quickly, and these shock waves seemed like the perfect stirring mechanism[13]

5.2.6   Nanotechnology

Nanotechnology is the science of the small with big potential. It is one of the most rapidly emerging key technologies. In recent years, noble metal nano-particles have been the subject focused research due to their unique electronic, mechanical, magnetic, optical, and chemical properties that are significantly different from those of bulk materials[13]. Therefore, metallic nano-particles have found use in more applications in different fields. Materials in the range of 1 NM-100 NM holds much interest because it is in this range that a number of newer properties become effective. The most widely used example of textile finishes by nanotechnology is of anti-microbial finishing. However, the use of textile finishing agents has been done on finishing of textiles with nano-particles as antibacterial compounds. Due to the increase in awareness about health and hygiene, people increasingly want their clothing to be hygienically fresh. Many of the anti-microbial agents available in the market are synthetic based and may not be environmentally friendly. Due to this, many of the consumers are opting for herbal anti-microbial finishes for textiles[15]. It must be ensured that these substances are not only permanently effective but also that they are compatible with the skin and environment.

5.2.7   Enzymatic finishing

The term Enzyme is derived from the Greek word "Enzymos" which means "in the cell or ferments". They are complex protein ferments secreted by living organisms and are believed to be as old as life itself. Enzymes can be isolated from animals, plants, or microbial origin where they play an important role in the function of cells and can be considered as living catalysts. They are able to grow and multiply themselves independent of the parent bodies under favorable conditions of time, temperature, concentration, pH, nutrients, salt, and in the absence of antiseptics and other inhibitors of enzyme action[16].

5.2.8   Sources of enzymes

Pancreatic enzymes are prepared from slaughterhouse waste such as the pancreas, clotted blood, liver, etc, whereas malt extracts are made from germinated barley. Bacterial enzymes are produced by growing cultures of certain micro-organisms in sterilized wort, providing an excellent supply of enzymes[17-18].

5.2.8.1   Use of enzymes in textiles

Textile industries use various chemical agents in their different processes like desizing, cotton softening, denim washing, silk degumming, etc. These chemicals after their use, cause pollution in the effluents; some of them are corrosive which could damage equipment and the fabric itself. With the introduction of enzymatic processes in textiles, the scenarios have changed in recent times, ensuring eco-friendly production and are successfully used in various textile processes like pre-treatment, dyeing, and finishing. Enzymes being natural products are completely bio-degradable and accomplish their work quietly and efficiently without leaving any pollutant behind. In addition, the process would operate at relatively low temperatures and atmospheric pressure with little by-product formation[18]. Crease resistance of cotton garments can be improved by enzyme-catalyzed cross-linking reactions at room temperature, e.g. Lipase class of enzymes can be used to promote cross-linking reactions[19].

5.2.9   Enzyme wash

Cellulase enzymes are natural proteins which are used in denim garment processing to get stone wash look onto the denim garments without using stones or by reducing the use of pumice stone. Cellulose attacks primarily on the surface of the cellulose fiber, leaving the interior of the fiber as it is, by removing the indigo present in the surface layer of fiber.

Cellulase enzymes are classified into two classes:
1. **Acid Cellulase:** It works best in the pH range of 4.5-5.5 and exhibits optimum activity at 50 deg C.
2. **Neutral cellulase:** It works best at pH 6 however its activity is not adversely affected in the range of pH 6-8 and show maximum activity at 55 deg C[20].

5.2.10   Microencapsulation method

Microencapsulation is one of the most popular methods of getting functional finishes on textiles. Microencapsulation is a micro-packaging technique involving deposition of the thin polymeric coating on small particles of solid or liquid. This process is more advantageous to conventional process in terms of economy, energy saving, eco-friendliness and controlled release of substances. The anti-bacterial agents reside in colloidal suspension with the amorphous zone of the polymeric binder so that a reservoir of the agent is present in solid/solution within the polymer matrix[21-22].

5.3   Eco friendly anti-microbial finishing using bio active agents

In the present world scenario, quality requirements require functionality and long service life along with environmentally friendly production process. Regular synthetic based anti-microbial agents are tri-closan, metal, and its salts, organometallics, phenols, quaternary ammonium compounds etc[23]. These synthetic anti-microbial agents are effective against microbes but provide some side effects, act on good micro-organisms and create water pollution, which results in the high demand of research on eco-friendly anti-microbial agents based on natural products is popular with respect to anti-microbial finishing. Some of the natural anti-microbial agents are chitosan[24], natural dyes[25-27], neem extract; herbal products such as aloe vera, tea tree oil, Eucalyptus oil and tulsi leaf extracts, etc. different kinds of active ingredients were extracted from the natural products. In general, the medicinal plants have active anti-microbial ingredients. Natural anti-microbial agents have lower side effects and lesser cost compared to synthetic agents and environment friendly. Among the entire natural anti-microbial agents' plant products covering major segment. Some of the natural anti-microbial agents are described below.

5.3.1   Chitosan

Chitosan [poly- (1-4) – 2 – amino 2 – deoxy – D –glucan], a deacetylated derivative of chitin is a natural, non – toxic, microbial resistant and biodegradable polymer. Chitin is one of the most abundant polysaccharides found in nature, derived from marine shells and mollusks[28]. The process of deacetylation involves the removal of acetyl groups from the molecular chain of chitin, leaving behind a high degree chemically reactive amino group. Chitin is made up of a linear chain of acetylglucosamine groups while chitosan is obtained by removing enough acetyl groups leaving reactive - NH_2 group in the molecule. Due to this chitosan, have two advantages over chitin. The first advantage is that chitosan is readily dissolved in dilute acetic acid, unlike chitin, which requires toxic solvents. The second advantage is that chitosan possesses free amine groups, which are the active sites for many chemical reactions[29]. Anti-fungal and antimicrobial properties of chitosan are believed to originate from its polycationic nature that can bind negatively charged residues of macromolecules at the cell surface of bacteria and then inhibit the growth of bacterial[30]. Chitosan and its derivatives have received a lot of attention as antimicrobial agents for use in textiles.

5.3.2   Sericin

Silk sericin is a natural protein derived from the silkworm. Sericin constitutes 25– 30% of silk protein. Most of the sericin is removed during raw silk production. Sericin is a biomolecule of great value as it is antibacterial, UV resistant, oxidative resistant and moisturizing properties. The recovery of sericin from degumming liquor reduces environmental pollution and as well, it has a lot of commercial applications in creams and shampoos as moisturizing agents and can be as used as bio material in the textile surface treatment[31]. Functional properties such as hygroscopic and little improvement in anti-bacterial property of polyester were improved using sericin application[32].

5.3.3   Neem extract

Neem is an evergreen tree, which is recognized as one of the effective source of insect control, antimicrobial and medicinal properties[33]. More than 300 different active compounds have been reported from the different parts of the tree. The neem extracts have been widely used in herbal pesticide formulation because of its pest repellent properties, which inhibit the growth of both gram-positive and gram-negative bacteria[34]. Application of neem as an antimicrobial agent in textiles is reported and few patents were filed based on the application of neem oil using micro encapsulation technique[35].

5.3.4   Natural dyes

Many natural dyes derived from plants are having antimicrobial properties. Pomegranate and many other natural dyes act as good antimicrobial agents due to the presence of tannins. Gupta et.al studied antimicrobial properties of eleven natural dyes against gram-negative and gram-positive bacteria. In addition, they found that the antimicrobial efficacy of a dye varies in the solution and in the textile substrate[36].

5.4   Eco-friendly finishing and washing effects of garments

* Bio-polishing
* Sandblasting
* Hydrophilic finish
* Wrinkle free treatment
* Water/oil repellent finish
* UV protection
* Soil release finish

5.4.1   Bio-polishing

To produce this kind of effect, the cellulose enzymes called GMO's (Genetically Modified Organisms) are produced. Before the making of GMO's, Acid cellulose was used for this effect. The bio-polishing process targets the removal of the small fiber ends protruding from the yarn surface and thereby reduces the hairiness or fuzz of the fabrics. The hydrolysis action of the enzyme weakens the protruding fibers to the extent that a small physical abrasion force is sufficient to break and remove them. Bio-polishing can be accomplished at any time during wet processing but is most convenient performed after bleaching.

It can be done in both continuous or batch processes. However, continuous processes require some incubation time for enzymatic degradation to take place. Removing the fuzz makes the color brighter, the fabric texture more obvious, and reduces pilling. Unfortunately, the treatment also reduces the fabric strength. Smoother yarns also increase fabric softness, appearance, and feel. Since it is an additional process, the bio-polished garments may cost slightly more. Next time you buy apparel, look for the label "Bio-Polished"[37].

5.4.2   Sand blasting

Sand blasting technique is based on blasting an abrasive material in granular, powdered, or other form through a nozzle at very high speed and pressure onto specific areas of the garment surface to be treated to give the desired distressed/ abraded/used look[38].

5.4.3   Hydrophilic finish

Ever since synthetic fibers became popular for clothing purposes, there has been the desire for a finish to change the hydrophobic character of these fibers. The main reason was to improve the wearing comfort. Hence the necessity to improve synthetic fibers with regard to their absorbency. One area of textile finishing where improving the absorbency is still one of the main considerations are sportswear, some of which are also made with a functional jersey with hydrophobic synthetic fibers on the inside and hydrophilic cellulosic fibers on the outside. The mode of action consists of the finest fibrilled microfibers (PES, PA or PP) transporting the moisture rapidly from the skin through the capillary interstices to the absorbent outer layer. In this way, the textile layer of synthetic fibers next to the skin remains dry. After dyeing the hydrophobic synthetic fibers usually exhibits no absorbency. Only after the application of a suitable hydrophilic agent can the material fulfill its function. This significantly increases the speed at which the moisture has spread to the hydrophilic outer layer and thus considerably accelerates drying[39].

5.4.4   Wrikle free treatment

By applying resins, it is possible to improve specific properties of cellulosic fibers. Examples of this kind are the improvement in crease recovery, dimensional stability, non-iron, reduced pilling and particularly with knit goods an improved appearance after several washes. For successful resin finishing, it is essential that the goods are well prepared and the recipes and processes are adhered to and monitored exactly.

The wrinkle-free treatment package comprises of low formaldehyde resin, silicones, and polyethylene emulsion. This treatment involves the chemical application of the elements comprising of this package through a cross-linking effect that prevents the formation of creases and wrinkles which result in easy to iron fabric. Resins do however also have several effects on the fibers. Resins reduce the (tear) strength of cotton. The extent of the loss depends on a wide variety of factors such as

- Amount and type of resin applied
- Amount and type of catalyst
- Curing conditions
- Quality of cotton
- Processes preceding finishing

Tensile strength losses up to 30-45% could be expected. For the so-called non-iron finishes, it is therefore often necessary to use qualities with a higher initial strength than for normal softening finishes. In this connection, it should be mentioned that the tensile strengths are not normally improved by the additives and softeners used[40].

5.4.5   Water-repellant finish

This finish gives hydrophobic features to the substrate. There are three main product groups for this finish

- Metal salt paraffin dispersion
- Polysiloxane

When finishing with these products, the surface of the goods must be covered with molecules in such a way that their hydrophobic radicals are ideally positioned as parallel as possible facing outwards. Aluminium salt paraffin dispersions are positively charged products due to the trivalent aluminium salt. This produces a counter polar charge on the fiber surface, which is significant for the adsorption of the product.

After drying, the fat radicals form a so-called "brush" perpendicular to the fiber surface, which prevents water drops from penetrating into the fiber. Polysiloxanes form a fiber-encircling silicone film with methyl group's perpendicular to the surface. The oxygen atoms are facing towards the fiber. The film formation and direction of the methyl groups are responsible for the hydrophobic properties of the finish[41].

5.4.6   UV protection

Fabric treated with UV absorbers ensures that the clothes deflect the harmful ultraviolet rays of the sun, reducing a person's UVR exposure and protecting the skin from potential damage. The extent of skin protection required by different types of human skin depends on UV radiation intensity and geographical location, time of day, and season. This protection is expressed as SPF (Sun Protection Factor), higher the SPF value better is the protection against UV radiation.

The SPF value of textile depends on fiber type, the fabric construction (porosity and thickness), and the finish. It means that transmission, absorption and reflectance nature of textile influence SPF value. It provides vital information about the fabric's sun protection ability. By using UV absorbers, exposure of the textile to UV lights is reduced on the one hand as well as the intensity of the transmitted UV light on the other. Good skin protection is achieved by the textile itself with sufficient weight of the fabric. A UV absorber can be applied either during fiber manufacture or in the final finish, which also offers the same degree of protection[36].

5.4.7   Soil release finish

Soil release finish facilitates removal of waterborne and oil stains from fabrics such as polyester and cotton blends and fabrics treated for the durable press, which usually show some resistance to stain removal by normal cleaning processes. This finish is especially suitable for sportswear, underwear, uniforms, and work wear, etc. These finishes are provided by nano-particles, which have high surface energy, the finish being durable up to 50 washes[42].

5.5   Research works related to eco – friendly finish

Sarkar et al. (2003) studied the use of clove, neem, tulsi and karanja oil on cotton fabric for their anti-bacterial property and observed that clove and neem oil show the good anti-bacterial property. 1% clove oil with KVSI (DMDHEU- dimethyloldihydroxyethylene ureabased inbuilt catalyst) exhibit excellent results[43].

Joshi et al. (2009) studied the application of various natural herbal extracts, chitosan, and natural dyes on textiles and analysed their impact on their anti-microbial behaviour. They found that the durability, shelf life, and anti-microbial efficiency are good with these agents[44].

Sathianarayanan et al. (2010) studied the application of Ocimum sanctum (tulsi leaf) and rind of Punicagranatum (pomegranate) on cotton fabric. The finish application was done by using the direct application method, micro-encapsulation, resin cross-linking, and their combinations. It has been found that the anti-microbial property is good with these elements[45].

Specos et al. (2010) studied the mosquito repellent effect of cotton fabrics treated with citronella oil through usage by humans. The subjects wore gloves made from mosquito repellent finished fabric and observed the effect by keeping their hands inside the box with Aedes aegypti mosquitoes. The fabric treated with microencapsulated citronella showed better results as compared to the ethanol solution of essential oils. It was concluded that microencapsulation of finish by pad-dry-cure method gives good results[46].

Chandrasekaran .et al. (2012) studied the application of 16 medicinal herbs such as neem, turmeric, basil, sandalwood, etc., on cotton for treatment of seven different diseases such as dermatitis, psoriasis, asthma, liver disorders, headache, joint pain, sinus/cold and found that there was a correlation between the antibacterial activity and the healing property of fabric[47]. Sumithra and Raaja (2012) studied the application of herbal extracts of Ricinuscommunis, Sennaauriculata, and Euphorbia hirta on 100% cotton denim fabric. It was found that the best combination of herbal extracts .i.e. Ricinuscommunis, Sennaauriculata, and Euphorbia hirta was (1:2:3) when applied through an exhaustion method on fabric. The finish lasted for 30 industrial washes and showed good anti-microbial resistance against microbes[48].

Sundarajan and Rukmani (2012) studied the antibacterial properties of limonene after applying it to cotton through microencapsulation by using gum acacia as wall material. It was found that limonene microcapsules were fixed to the fabric by using crosslinker citric acid, which formed covalent bonds with the fabric due to which the fabric possesses very good antibacterial property even after five washing cycles. The durability of the finish can be increased by using a fixing agent so that its wash durability can be increased[49]. Aparna and Krishnaveni (2014) studied the application of Aloe Vera gel on 100% cotton knit single jersey through micro encapsulation technique. Microcapsules were formed using herbal extracts as core and gum acacia as wall material. These were applied on fabric by a simple pad-dry-cure method, which was further used in the treatment of skin diseases. It can be concluded that herbal cloth can be made by microencapsulation of essential oils through padding[50].

Geethadevi and Maheshwari (2014) studied the application of essential oils like thyme oil, cypress oil, and grapefruit oils in combination with natural gums like sodium alginate, acaciaarabica, and moringaoleifera by using microencapsulation technique as essential oils in the core material. These oils were mixed in different ratio and final ratio selected was 2:1:1. The three natural gum materials were used as shell materials for good durability. It was found that Moringa Oleifera finished fabric showed good results for UV protection, mosquito repellence, and no allergic reaction was found on skin[51]. West and Hitchcock (2014) studied the application of essential oils on textiles with the usage of different methods like Pad-Dry-Cure, Microencapsulation, and Mixed methods. These aromatic oils have medicinal and anti-microbial properties benefiting humankind. They are eco-friendly and can be used as medicinal textiles. It was found that microencapsulation of finish yielded good results with pad-dry-cure method because the finish were interlaced between the spaces in the fibre and was not present superficially[52]. Bano (2014) studied the application of mosquito repellent finish on cotton knitted fabric by using a commercial binder and replacing it with natural binder chitosan. It was found that chitosan was more durable as compared to a commercial binder in terms of repellence and wash durability[53]. Javid et al. (2014) studied the microencapsulation of essential oils with chitosan in the presence of cross-linking agent dihydroxy ethylene urea and a bio surfactant with respect to their size, morphology, and stability and found that anti-bacterial activity increases with increase in the concentration of chitosan and essential oils[54].

5.6   Conclusion

- A finish is a treatment given to a piece of fabric, to change its appearance, handling/touch or its purpose is to make the fabric more suitable for its end use.

- Finish includes any general treatment given to clean and iron fabrics and create exclusive variations of them by using chemical treatments, dyeing, and printing to make fabric attractive and appealing.

- Mainly need for finishing, improve the appearance of fabric, improve the touch or feel of the fabric, improve the draping ability of lightweight fabric, produce a variety of fabrics through dyeing and printing. Make fabric suitable for specific end uses and enhance sale appeal.

- Enzymatic finishing, plasma technique, microencapsulation, and nanotechnology are also very good processes for giving eco-friendly finishes.

- The plasma technique proved very effective as it consumes low energy and chemicals and there is no problem with the disposal of waste. Enzymes being natural products are completely biodegradable and leave no pollutant behind. With the introduction of these processes in textile processing, the scenario has changed in recent times ensuring eco-friendly products.

- With the increase in pollution, the environment related problems are increasing day by day. Moreover, the textile industry holds a major position in this environmental pollution. Therefore, it is a moral duty of every individual to adopt such technologies that imparts in the well-being of the environment, which in turn will be the well-being of living organisms too.

5.7   References

1. Challa. L (2013), Impact of Textile and clothing industry on environment; approach towards eco-friendly textiles, article retrieved from http://fibre2fashion.com.

2. Chandrashekharan K., Ramchandran T., Vighneshwaran C (July 2012), Effects of medicinal herb treated garments on selected diseases, Indian Journal of traditional knowledge, vol. 11 (3), 493 – 498.

3. Mahesh S., Manjunatha R &Vijayakumar G (2011). Studies on antimicrobial textile finish using certain plant natural products. International conference on advances in biotechnology and pharmaceutical science (ICABPS 2011) Dec., 2011.

4. Srinivasan G., (2013), N9 Pure silver™Antimicrobial for hygieneic textile applications, Asian Textile journal, 52 – 58.

5. Adivarekar R.V.., Kannongo N., Nerurkar M., Khurana N. (2011). Application of herbal extracts for anti-microbial property, Journal of the textile association, 324 – 330.

6. Chakraborty, Pal R., Kaur R. (2006), Plasma treatment of textiles, Asian textile journal, 67 – 75.

7. Sudha S., Giridev V.R., Neelakandan r., (2006), Plasma application in textiles – an overview, journal of the textile association, 25 – 29.

8. http:/www.4thstate.com

9. ParvinzadehGashti, Mazeyar. (2012). Surface modification of synthetic fibers to improve performance: Recent approaches. Global Journal of Physical Chemistry. 3. 1-10.

10. Doshi, Rashesh and VinodShelke. 2001. "Enzymes in Textile Industry – An Environment-Friendly Approach." Indian Journal of Fibre and Textile Research 26(1–2):202–5.

11. AbouElmaaty, Tarek&Abd El-aziz, Eman. (2017). Supercritical carbon dioxide as a green media in textile dyeing: A review. Textile Research Journal. 004051751769763. 10.1177/0040517517697639

12. Vajnhandl, S., & Le Marechal, A. M. (2005). Ultrasound in textile dyeing and the decolouration/mineralization of textile dyes. Dyes and Pigments, 65(2), 89-101.

13. Duron N., et al., (2007), journal of bio-medical nanotechnology, 3, 203 – 208.

14. Kavita T., Padmashwini R., Giridev V R., Neelakantan R. (2006), Synthetic fibres, 4 – 15.

15. Malik T., et al., http://www.fibre2fashion.com.

16. Boyer P D., (1959)., Handbook of Enzymes, Ed. P.D. Boyer, Vol 1, 136.

17. Shah D.L., (1990), Man-made textiles in India, 33, 426.

18. Mitra A., Saylee P., rathi C.L., (1995), Chemical weekly, 12, 155.

19. Li. Y., Hardin R (1997), Text. Chem, Color. 29 (8), 71.

20. Duran, N., & Duran, M. (2000). Enzyme applications in the textile industry. Review of Progress in Coloration and Related Topics, 30(1), 41-44.

21. Sivaramkrishnan C.N. (2007), Colourage, 36 – 38.

22. Thilagavathi G., Krishna Bala S., (2007), Indian Journal of Fibre and textile Research, 32, 351 – 354.

23. Purwar, R., & Joshi, M. (2004). Recent Developments in Antimicrobial Finishing of Textiles--A Review. AATCC review, 4(3).

24. Lim, S. H., & Hudson, S. M. (2003). Review of chitosan and its derivatives as antimicrobial agents and their uses as textile chemicals. Journal of Macromolecular Science, Part C: Polymer Reviews, 43(2), 223-269.

25. Gupta, D., Khare, S. K., &Laha, A. (2004). Antimicrobial properties of natural dyes against Gram-negative bacteria. Coloration Technology, 120(4), 167-171.

26. Singh, R., Jain, A., Panwar, S., Gupta, D., &Khare, S. K. (2005). Antimicrobial activity of some natural dyes. Dyes and pigments, 66(2), 99-102.

27. Gupta, D., Jain, A., &Panwar, S. (2005). Anti-UV and anti-microbial properties of some natural dyes on cotton. Indian Journal of Fibre& Textile Research, 30(2), 190-195.

28. Dutta, P. K., Ravikumar, M. N. V., &Dutta, J. (2002). Chitin and chitosan for versatile applications. Journal of Macromolecular Science, Part C: Polymer Reviews, 42(3), 307-354.

29. Tsigos, I., Martinou, A., Kafetzopoulos, D., &Bouriotis, V. (2000). Chitin deacetylases: new, versatile tools in biotechnology. Trends in biotechnology, 18(7), 305-312.

30. Jeon, Y. J., Park, P. J., & Kim, S. K. (2001). Antimicrobial effect of chitooligosaccharides produced by bioreactor. Carbohydrate polymers, 44(1), 71-76.

31. Kavitha, T., Padmashwini, R., Swarna, A., Dev, V. R., Neelakandan, R., & Kumar, M. S. (2007). Effect of chitosan treatment on the properties of turmeric dyed cotton yarn.

32. Yamada, H., & Nomura, M. (1998). Fibrous article for contact with skin. Japan patent.

33. Singh R P, Chari M S, Raheja A K & Kraus W, Neem and Environment, Oxford & IBH publishing, New Delhi, 1996.

34. Elakovich, S. D. (1996). The Neem Tree: Source of Unique Natural Products for Integrated Pest Management, Medicinal, Industrial, & Other Purposes Edited by Heinrich Schmutterer (Giessen U., FGR). VCH: New York. 1995. xxi+ 680 pp. $125.00. ISBN 3-527-30054-6.

35. Nathalie C., WO Pat 03002807 (2003)

36. Saravanan, D. (2007). UV protection textile materials. AUTEX Research Journal, 7(1), 53-62.

37. Aly, A. S., Moustafa, A. B., &Hebeish, A. (2004). Bio-technological treatment of cellulosic textiles. Journal of Cleaner Production, 12(7), 697-705

38. Tarhan, M., &Sarıışık, M. (2009). A comparison among performance characteristics of various denim fading processes. Textile Research Journal, 79(4), 301-309.

39. Linford, M., Lau, R., Soane, D., Millward, D., Green, E., & Ware, W. (2003). U.S.

Patent Application No. 10/136,191.

40. Hashem, M., Ibrahim, N. A., El-Shafei, A., Refaie, R., & Hauser, P. (2009). An eco-friendly–novel approach for attaining wrinkle–free/soft-hand cotton fabric. Carbohydrate Polymers, 78(4), 690-703.

41. Ceria, A., & Hauser, P. J. (2010). Atmospheric plasma treatment to improve durability of a water and oil repellent finishing for acrylic fabrics. Surface and Coatings Technology, 204(9-10), 1535-1541.

42. Paul, R. (2014). Functional finishes for textiles: an overview. Functional Finishes for Textiles, Improving Comfort, Performance and Protection, 1-14.

43. Sarkar, R.K., Purushottam, D. and Chauhan, P.D. (2003). Bacteria-resist finish on cotton fabrics usingnatural herbal extracts. Indian J. Fibre Text Res., 28 (9), : 322-331.

44. Joshi, M., Wazed, Ali S., Purwar, R and Rajendran, S. (2009). Ecofriendly antimicrobial finishing oftextiles using bioactive agents based on natural products. Indian J. Fibre Text Res, 34 (9): 295-304.

45. Sathianarayanan, M.P., Bhat, N.V., Kokate, S.S. and Walunj, V.E. (2010). Antibacterial finish for cottonfabric from herbal products, Indian J. Fibre & Textile Res., 35 : 50-58

46. Specos, Moro M.M., Garcia, J.J., Tornesello, J., Marino, P., Vecchia, Della M., TesorieroDefain, M.V.andHermida, L.G. (2010), Microencapsulated citronella oil for mosquito repellent finishing ofcotton textiles. Transactions Royal Soc. Tropical Med. & Hygiene, 104 : 653-658

47. Chandrasekeran, K., Ramachandran, T. and Vigneswaran, C. (2012). Effect of medicinal herb extractstreated garments on selected diseases, Indian J. Traditional Knowledge, 11(3) : 493-498.

48. Sumithra, M. and Raaja, Vasugi N. (2012). Micro-encapsulation and nano-encapsulation of denimfabrics with herbal extracts. Indian J. Fibre & Textile Res., 37 : 321-325.

49. Sundrarajan, M. and Rukmani, A. (2013). Durable Anti-Bacterial finishing on cotton by impregnationof Limonene Microcapsules. American Scientific Publishers, Advanced Chemistry Letters.1 :40-43

50. Krishnaveni, V. and Aparna, B. (2014). Microencapsulation of copper enriched Aloe gel curativegarment for atopic dermatitis. Indian J. Traditional Knowledge, 13(4) : 795-803.

51. Geethadevi, R. and Maheshwari, V. (2015). Long-lasting UV protection and mosquito repellent finishon bamboo/tencel blended fabric with microencapsulated essential oil. Indian J. Fibre & Textile Res., 40 (6) : 175- 179.

52. West, A..J and Annett-Hitchcock, K.E. (2014). A critical review of aroma therapeutic applications fortextiles. J. Textile & Apparel, Technol. & Mgmt., 9 (1) : 1-13.

53. Bano, R. (2014). Use of chitosan in mosquito repellent finishing for cotton textiles. J. Textile Sci. Eng.,4 (5): 1-3.

54. Javid, Amjed, Raza Ali Zulfiqar, Hussain, Tanveer and Rehman, Asma (2014). Chitosan microencapsulation of various essential oils to enhance the functional properties of cottonfabrics. J. Microencapsulation (Research Gate), pp 1-8.

6

Nanotechnology – A novel route towards finishing of apparels

S. Manjula

Associate Professor and Head, Department of Costume Design and Fashion,
Kongu Arts and Science College (Autonomous),, Erode-638 107. Email: manjulalokgu@gmail.com

Abstract : Advances in nanotechnology have created enormous opportunities in the realm of apparel finishing sector, which has an ever-increasing demand for improved multifunctional applications. Nanoparticles owing to their large surface area-to-volume ratio and high surface energy, have spurred significant developments in imparting novel functionalities such as microbial resistance, anti-static effect, water repellency, flame retardancy, wrinkle resistance, UV-protection and self-cleaning property to textiles. Nanofinishing helps to achieve uniform distribution of required functionalities with a limited amount of additives. The quest for enhanced performance and efficiency has opened up new routes in nanotechnology for the production of functional apparels. This review focuses on the application of various nanoparticles for specific functionalization of textiles and apparels.

Keywords : Nanotechnology, nanoparticles, self-cleaning, anti-microbial, flame retardancy, water repellency, UV- protection, anti-static, wrinkle resistance.

6.1   Introduction

Never in the history of Science and Technology has so much euphoria been created among scientists of varied disciplines by the magical word 'nanotechnology'. Today's nanotechnology has already crossed several new and unforeseen frontiers. Nanotechnology is pursued vigorously for discovering new phenomena and for developing new functionalized high-performance materials. Nanotechnology is indeed projected by some as a solution to almost all problems of humanity.[1]

Nanotechnology is science, engineering, and technology conducted at the nanoscale, which ranges from 1 to 100 nanometres. The concept of nanotechnology evolved in a talk by physicist Richard Feyman on the title 'There is plenty of room at the bottom' at an American Physical Society meeting held at the California Institute of Technology (CalTech) on December 29, 1959. He described a process in which scientists would be able to manipulate and control individual atoms and molecules. After a decade, Professor Norio Taniguchi coined the term 'nanotechnology'. Modern nanotechnology began after 1981, with the development of the scanning tunneling microscope that could see individual atoms. One nanometer is a billionth of a meter or 10^{-9} of a meter, tens of thousands of times smaller than the width of human hair.[2]

The use of nanomaterials and nanotechnology-based processes is growing at a tremendous rate in all fields of science and technology. The textile industry is also experiencing the benefits of nanotechnology in its diverse field of applications. Textile-based nano products starting from nanocomposite fibers, nanofibers to intelligent high-performance polymeric nanocoatings are getting their way not only in high performance advanced applications but also successfully being used in conventional textiles to impart new functionality and improved performance. Greater repeatability, reliability, and robustness are the main advantages of nanotechnological advancements in textiles. Nanoparticles are the fundamental building blocks for nanotechnology applications. Nanoparticle application during conventional textile processing techniques, such as finishing, coating, and dyeing, enhances the product performance manifold and imparts hitherto unachieved functionality.[3]

Nanotechnology has opened immense possibilities in textile finishing area resulting in innovative finishes as well as new application techniques. Particular emphasis is on making chemical finishing more controllable, durable and significantly enhancing the functionality by incorporating various nanoparticles or creating nanostructured surfaces. The unprecedented level of textile performances claimed for these nanofinishes such as stain resistance, antimicrobial, controlled hydrophilicity/hydrophobicity, antistatic, UV protective, wrinkle resistant abilities can be exploited for a range of applications in the apparel industry.[4]

6.2   Self cleaning finish

Nature has already developed an elegant approach that combines chemistry and physics to create super repellant surfaces as well as self-cleaning surfaces. *Lotus leaves* is the best example of self-cleaning surfaces. The concept of self-cleaning textiles is based on the lotus plant whose leaves are well known for their ability to self-clean by repelling water and dirt (Fig. 6.1). More recently, botany and nanotechnology have united to explore not only the beauty and cleanliness of the leaf, but also its lack of contamination and bacteria, despite its dwelling in dirty ponds. The lotus leaf has two levels of structure affecting this behavior - micro-scale bumps and nano-scale hair-like structures coupled with the leaf's waxy chemical composition. Based on the lotus leaf concept, scientists developed a new concept called 'Self-cleaning textile' the textile surface, which can be, cleaned itself without using any laundering action.[5]

Water particles are unable to seep through the textile.
This is called the "lotus-effect."

Figure 6.1: Lotus Leaf Effect

There are two principal ways of self-cleaning materials, namely hydrophobicity, and hydrophilicity. Both types of coating clean themselves with the action of water by rolling droplets for hydrophobic and sheeting water for hydrophilic that carries dirt away. Nevertheless, hydrophilic have an additional property, which can chemically break down the adsorbed dirt in sunlight through the help of photocatalyst which also known as hydrophilic photocatalytic coating.[6] Hydrophobic surface repels water with the properties of low wettability and contact angles more than 90°. The higher the contact angle the lower the value of adhesion on the surface and result in increased hydrophobicity. For contact angle, more than 150° the surface is termed as superhydrophobic. Hydrophobicity can also be regulated by the roughness factor. With rougher surfaces, the contact angle of water on it will increase and form bumps that trap air between water and the surface. The "Lotus Effect" was applied in this superhydrophobic mechanism. The lotus plant cleans itself by having a superhydrophobic leaf, which consists of microscopic bumps all across the leaf's surface that plays an important part in its water-repelling properties. A rough coating of nanoscopic wax crystals on these bumps further increases the effect. It allows the water droplets to roll across and removes the dirt away.[7]

Photocatalytic process is the acceleration of a photoreaction in the presence of a catalyst. This process will decompose the dirt molecules by utilizing the sunlight. By utilizing the photoreaction induced by photocatalyst, the organic contaminants will be degraded into air and water. The mechanism of photocatalytic reaction begins when a photocatalyst is irradiated by light, usually ultraviolet light (Fig. 6.2). Photocatalytic nanoparticles such as TiO_2

and ZnO have been used to produce hydrophilic surfaces with self-cleaning activity. Self-cleaning effect of a textile material can be obtained by a photo-catalytically active coating containing a photo-catalytically active oxide of a transition metal such as titanium dioxide (TiO_2). Due to light absorption in the near UV, electrons are hoisted from the energy level of the valence band of TiO_2 into that of the conductive band, thus leaving a positively charged hole in the valence band. Titanium dioxide can also destroy pathogens such as bacteria in the presence of sunlight by breaking down the cell walls of the microorganisms. Zinc oxide is also a photocatalyst, and its photocatalysis mechanism is similar to that of titanium dioxide.[8]

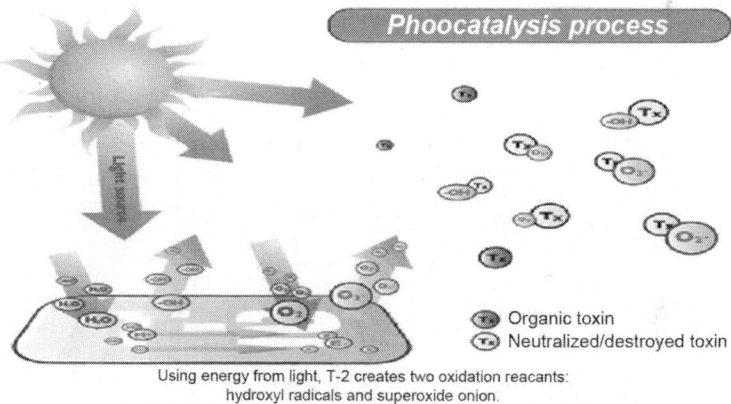

Figure 6.2: Mechanism of Titanium dioxide photocatalytic degradation process

6.3   Anti-microbial finish

With the growth in world population and the spread of diseases, the number of antibiotic resistant microorganisms is rising along with the occurrence of infections from these microorganisms. With this increase in health awareness, many people focused their attention on educating and protecting themselves against harmful pathogens. It soon became more important for antimicrobial finished textiles to protect the wearer from bacteria than it was to simply protect the garment from fiber degradation[9]. For imparting anti-bacterial properties, TiO_2, zinc oxide, and nano-sized silver are used. Metallic ions and metallic compounds display a certain degree of sterilizing effect. With the use of nano-sized particles, the number of particles per unit area is increased, and thus anti-bacterial effects can be maximised[10]. Fabrics treated with nano-TiO_2 could provide effective protection against bacteria and the discoloration of stains, due to the photocatalytic activity of nano-TiO_2. On the other hand, zinc oxide is also a photocatalyst, and the photocatalysis mechanism is similar to that of titanium dioxide; only the band gap is different from titanium dioxide. Nano-ZnO provides effective photocatalytic properties once it is illuminated by light, and so it is employed to impart anti-bacterial properties to textiles.[11]

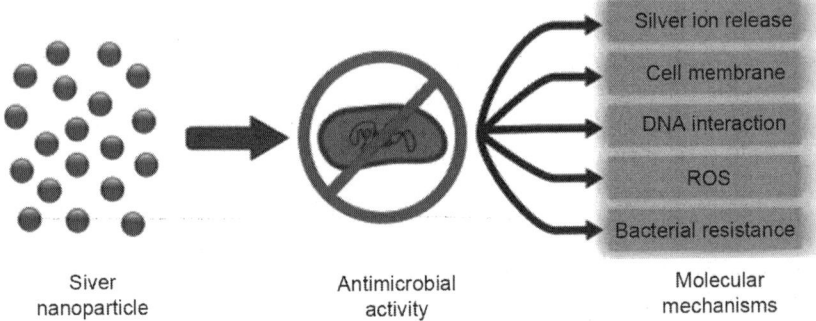

Siver
nanoparticle

Antimicrobial
activity

Molecular
mechanisms

Figure 6.3: Mechanism of action of silver nanoparticles

Compared to other salts, the efficient antimicrobial activity of silver nanoparticles is due to their extremely large surface area that provides more contact with microorganisms (Fig. 6.3). Some literature shows that there exists an electrostatic attraction between the positively charged nanoparticles and negatively charged bacterial cells. These nanoparticles get attached and penetrate into the cell membrane causing damage to the cells by attacking the respiratory chain and cell division leading to cell death.[12]

6.4   Flame retardant finish

Flame retardant materials do not continue burning the on removal of an ignition source, although there may be some changes in the property of the material on exposure to flame. An inherently flame resistant fiber may stop combustion by any of the following methods: heat may be removed or cooling applied; the increase in pyrolysis temperature of a material, making the material heat resistant; to prevent evaporation, that is to form non-volatile compounds in-situ called char. Meanwhile, aqueous dispersions of nanosized antimony pentoxide can be used with halogenated flame retardants for flame retardant treatments of materials such as textiles, nonwovens and flame retardant coatings. Nanoelements like multi-layered silicates appear to act as an excellent insulator and thus improve flame retardancy.[13]Colloidal antimony pentoxide has been offered as fine particle dispersion, for use as a flame retardant synergist with halogenated flame-retardants (the ratio of halogen to antimony is 5:1 to 2:1). Nano antimony pentoxide is used with halogenated flame-retardants for a flame retardant finish to the garments.[14] Cotton fabrics with good water–oil-repellence and flame-retardant properties with relatively durable properties could be produced using nanosols containing guanidine dihydrogen phosphate and urea, together with tetraethylorthosilicate and hexadecyltrimethoxysilane as precursors.[15]

6.5   Water repellent finish

Water-proof breathable fabrics prevent water from penetrating into the fibers but are breathable to allow water vapor to diffuse through fibers. Woven fabric, microporous membrane, coating, and smart breathable fabric are the different classifications of breathable fabrics. The first waterproof breathable fabric was prepared by densely woven fabric or polymeric and resin coating. However, nanotechnology has brought newer opportunities to develop water-proof breathable fabric.[16] Most successful developments in this regard can be attributed to a US-based company, Nano-Tex™(Fig. 6.4).

Figure 6.4: Water droplets on Nano-Tex fabrics

Their trademark *Nano-Pel* technology utilizes the concept of surface engineering and develops hydrophobic fabric surfaces that are capable of repelling liquids and resisting stains, while complementing the other desirable fabric attributes, such as breathability, softness, and comfort. In these fabrics, the water-repellent property of a fabric is improved by creating nanostructures called nanowhiskers, which are hydrocarbons and are 1/1000 of the size of a typical cotton fibre, that are added to the fabric to create a peach fuzz effect. Tiny whiskers aligned by proprietary "spines" are designed to repel liquids and are attached to the fibers utilizing molecular "hooks". The spaces between the whiskers on the fabric are smaller than the typical drop of water but still larger than water molecules; water thus remains on the top of the whiskers and above the surface of the fabric. Since the attached whiskers are of nanoscale size, they do not affect the hand, the breathability of the fabric and can withstand 50 home launderings.[17, 18]

Apart from Nano-Tex, a Swiss-based textile company Schoeller developed the NanoSphere (Fig. 6.5) to make water-repellent fabrics. NanoSphere impregnation involves a three-dimensional surface structure with gel-forming additives, which repel water and prevent dirt particles from attaching themselves. The mechanism is similar to the lotus effect occurring in nature.[19]

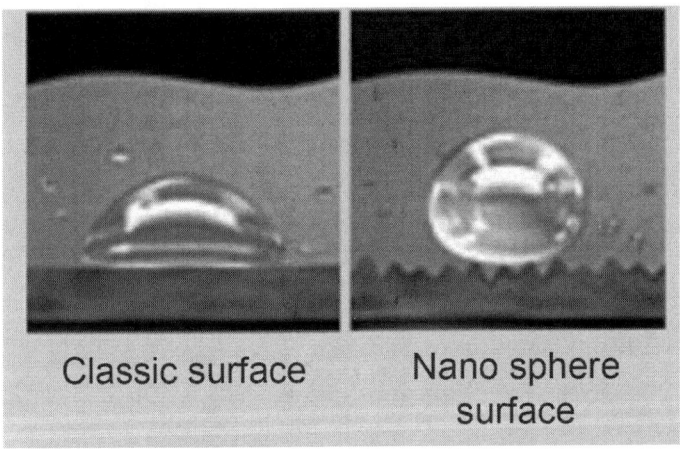

Figure 6.5: Schoeller Nanosphere Technology

Generally, hydrophobic finishes lower the surface energy and can give a maximum water contact angle of roughly 120□. This type of finish cannot be obtained by a surface coating. Superhydrophobicity increases with an increase in surface roughness which provides larger geometric area. The roughened surface generally takes the form of a substrate membrane with a multiplicity of microscale to nanoscale projections or cavities. Water repellency of rough surface is due to the air enclosed between the gaps in the surface. This enlarges the air/water interface while minimizing the solid/water interface. In this situation, spreading does not occur and the water forms a spherical droplet. SiO_2, Al_2O_3 nanoparticles are mainly used for super water repellent finishes.[20] On the other hand, the hydrophobic property is imparted by coating the fabric with a thin nanoparticulate plasma film. It has been demonstrated that by combining the nanoparticles of hydroxylapatite, TiO_2, ZnO, and Fe_7O_3 with other organic and inorganic substances, the audio frequency plasma of fluorocarbon chemical when applied onto a cotton fabric deposits a nanoparticulate hydrophobic film thereby improving its water repellent property.[21]

6.6   Hydrophilic finish

The poor moisture absorption property of synthetic fabrics such as polyester and polyamides limits its applications in the apparel sector. The new range of hydrophilic nanofinishes 'Cotton Touch'™ and 'Coolest Comfort'™ commercialized by NanoTex, USA makes the synthetic fabric look and feel like cotton. "Nanotouch® gives durable cellulose wrapping over synthetic fibers such as polyester and polyamides. Cellulosic sheath and synthetic core together form a concentric structure to bring overall solutions to the drawbacks of synthetics such as static discharge, harsh handle, and glaring luster. It can also last 50 launderings and expected to eliminate the decline in demand for synthetic microfiber and broaden the use of synthetics to new applications. 'Nano Dry® Finish provides breakthrough moisture wicking to draw moisture away the from body while drying quickly. It improves the moisture absorption of polyamides and polyesters making them hydrophilic and comfortable. The main applications are in sportswear and close to body garments that require perspiration absorbency. The finish lasts 50 launderings.[22, 23]

6.7   UV protection finish

Textile materials can provide protection against the harmful effects of UV radiation if necessary attention is paid while engineering the basic design and structure of the material (Fig. 6.6). Inorganic UV-blockers are more preferable to organic UV-blockers as they are non-toxic and chemically stable under exposure to both high temperatures and UV. Inorganic UV-blockers are usually certain semiconductor oxides such as TiO_2, ZnO, SiO_2 and Al_2O_3. Among these semiconductor oxides, TiO_2 and ZnO are commonly used. It was determined that nano-sized titanium dioxide and zinc oxide were more efficient at absorbing and scattering UV radiation than the conventional size, and were thus better able to block UV. This is due to the fact that nanoparticles have a larger surface area per unit mass and volume than the conventional materials, leading to the increase of the effectiveness of blocking UV radiation.[24] Through the uniform distribution of nanoparticles on the fabric surface, the performance of UV-blocking agents can be efficiently increased.[25]

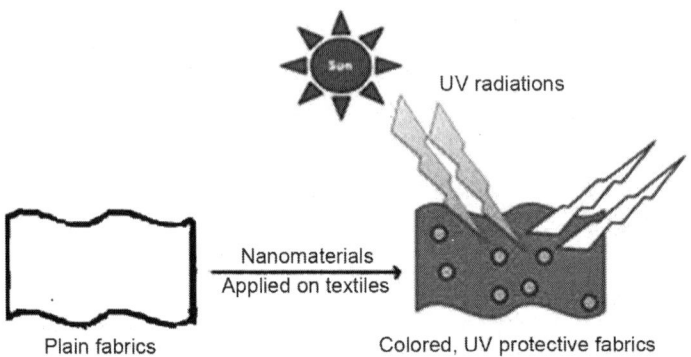

Figure 6.6: UV Protective Fabrics

6.8   Antistatic finish

An antistatic agent is a compound used for the treatment of materials or their surfaces in order to reduce or eliminate the buildup of static electricity generally caused by the triboelectric effect. As synthetic fibers provide poor anti-static properties, research work concerning the improvement of the anti-static properties of textiles by using nanotechnology has been at large. It was determined that nano-sized particles like titanium dioxide, zinc oxide whiskers, nano antimony-doped tin oxide, and silane nanosol could impart anti-static properties to synthetic fibers. Such materials help to effectively dissipate the static charge, which is accumulated on the fabric.[26]

6.9   Wrinkle resistant finish

To impart wrinkle resistance to fabric, the resin is commonly used in conventional methods. However, there are limitations to applying resin, as it causes a decrease in tensile strength of fibre, abrasion resistance, water absorbency, and dyeability, as well as breathability. To overcome the limitations of using resin, some researchers employed nano-titanium dioxide and nano-silica to improve the wrinkle resistance of cotton and silk respectively. Nano-titanium dioxide was employed with carboxylic acid as a catalyst under UV irradiation to catalyse the cross-linking reaction between the cellulose molecule and acid whereas nano-silica was applied with maleic anhydride as a catalyst.[27, 28]

6.10   Conclusion

The unique properties of nanomaterials have not only attracted scientists and research workers but also business holders due to its huge commercial potential. The application of nanoparticles for the development of multi-functional finishes to textiles has endless possibilities and there is no doubt that nanotechnology has reached its pinnacle in the textile-finishing sector by way of nano-coating. Nano-coating technology offers new functionalities such as self-cleaning surfaces, antimicrobial properties, controlled hydrophilicity, or hydrophobicity, protection against fire, UV radiation, etc. Since the performance is built-in on a nano or sub-micron scale, these features provide permanence to the targeted requirement of the fabrics without affecting its inherent nature such as breathability and handle. It is clear that nanotechnology is soon to revolutionize the apparel-finishing sector by offering exciting opportunities for future research and development.

6.11   References

1. Chopra, K.L., (2012), Nanotechnology: hope or hype?, Current Science, 102(10), 1364-1366.

2. (https://www.nano.gov/nanotech-101/what/definition)

3. Joshi,M and Bhattacharyya,A, (2011), Nanotechnology-a new route to high performance textiles, Textile Progress, 43(3), 155-233.

4. Joshi, M., (2005), Nanotechnology: Opportunities in Textiles, Indian Journal of Fiber and Textile Research, 30, 477-479.

5. https://www.technicaltextile.net/articles/self-cleaning-textile-an-overview-2646?no_redirect=true

6. Banerjee,S, Dionysiou,D.D and Pillai,S.C., (2015), Self-cleaning applications of TiO2 by photoinduced hydrophilicity and photocatalysis, *Applied Catalysis B*: Environmental, 176, 396–428.

7. Barthlott,W and Neinhuis,C, (1997), Self-cleaning of biological surfaces, Planta, 202, 1-8.

8. Jašková,V, Hochmannová,L and VytLasová,J, (2013), TiO2 and ZnO nanoparticles in photocatalytic and hygienic coatings, International Journal of Photoenergy, 1-6, Doi:10.1155/2013/795060.

9. Yadav,A, Prasad,V, Kathe,A.A, Raj,S, Yadav,D, Sundarmoorthy,C, Vigneshwaran,N, (2006), Functional finishing in cotton fabrics using zinc oxide nanoparticles, Bulletin of Materials Science, 29(6), 641–645.

10. Saito,M, (1993), Antibacterial, Deodorizing, and UV Absorbing Materials obtained with Zinc Oxide (ZnO) Coated Fabrics, Journal of Coated Fabrics, 23,150-164.

11. Chen,R.Q, (2002), Nanometer materials and health-care textiles, Dyestuff Industry, 39, 24-28.

12. Dahl,J.A, Maddux,B.L, Hutchison, J.E., (2007), Toward greener nanosynthesis, Chemical reviews, 107(6), 2228-2269.

13. Mahish, S.S, and Mukherjee,S (2006), Flame retardancy in textiles - Scope of nanotechnology, Asian Textile Journal, 15(12), 57-58.

14. Morshed, A.M.A, Ikbal, M.H, and Hasan, S.M.K, (2010), Impact of nanotechnology in the arena of textile apparel finishing, Bangladesh Textile Today, 3(3), 53-55.

15. Mete,G, (2016), Development of water-, oil-repellent and flame-retardant cotton fabrics , Journal of Textile Institute, 107(11), 1164-1177.

16. Mukhopadhyay,A and Midha,V.K, (2008), A review on designing the waterproof breathable fabrics part I: fundamental principles and designing aspects of breathable fabrics, Journal of Industrial Textiles, 37(3), 225–262.

17. Linford,M.R, Soane,D.S, and Offord,D.A, February 2005, US Patent 6,855,772 (to Nano Tex LLC.), 15.

18. Kathiervelu,S.S, (2003), Applications of nanotechnology in fibre finishing, Synthetic Fibres, 32, 20-22.

19. Russell,E, (2002), Nanotechnologies and the shrinking world of textiles, Textile Horizons, 9(10),7-9.

20. Bhushan,B, Nosonovsky,M and Jung,Y.C, (2007), Towards optimization of patterned superhydrophobic surfaces , Journal of the Royal Society Interface,4(15), 643-648.

21. Zhang,J, France,P, Radomyselskiy,A, Datta,S, Zhao,J and Ooij,W.V, (2003), Hydrophobic cotton fabric coated by a thin nanoparticulate plasma film, Journal of Applied Polymer Science, 88(6), 1473-1481.

22. Joshi,M, (2008), The Impact of Nanotechnology on Polyesters, Polyamides and other Textiles, Advances in Polyesters and Polyamides, Woodhead Publishing Ltd., Cambridge, UK.

23. https://www.nanotex.com/coolest-comfort/

24. Xin,J.H, Daoud,W.A, and Kong,Y.Y, (2004), A New Approach to UV-Blocking Treatment for Cotton Fabrics, Textile Research Journal, 74(2), 97-100.

25. Kathirvelu,S, Dsouza,L, Dhurai,B, (2009), UV protection finishing of textiles using ZnO nanoparticles, Indian Journal of Fiber and Textile Research, 34(3), 267–273.

26. Wong,Y.W.H, Yuen,C.W.M, Leung,M.Y.S, Ku,S.K.A, Lam,H.L.I, (2006), Selected applications of nanotechnology in textiles. AUTEX Research Journal, 6(1),1-8.

27. Wang,C.C, and Chen,C.C, (2005), Physical properties of crosslinked cellulose catalyzed with nano titanium dioxide, Journal of Applied Polymer Science, 97(6), 2450-2456.

28. Song,X.Q, Liu,A, Ji,C.T, and Li,H.T, (2001), The effect of nano-particle concentration and heating time in the anti-crinkle treatment of silk, Journal of Jilin Institute of Technology, 22, 24-27.

7

Study on enzymatic treatment for biosoftening finish on woven and knitted cotton fabrics

Gopalakrishnan Duraisamy & Dr. P. Kandhavadivu

Department of Fashion Technology, PSG College of Technology, Coimbatore – 641 004
Tamilnadu, India, E mail: dgk.fashion@psgtech.ac.in

Abstract : Enzymes are nothing but biological catalysts. They are either a fungal or a bacterial source. Enzymes are present in living organisms but are themselves not living organisms. Structurally they resemble proteins. These proteins greatly accelerate reactions, but cannot cause a reaction to occur that ordinary would not occur, however, in the absence of enzymes an extremely long time may be required for the same reactions to occur. Initially, enzymes were used only for the removal of starch but nowadays enzymes are widely used in textile wet processing. The present state of Scouring involves hot alkali and other auxiliaries affect the rupture of the primary wall. Hot alkaline scouring results in environmentally harsh chemicals bracing added to the textile effluent. Bleaching with conventional (peroxide or chlorine) are effective in bleaching and degrading colourants and other impurities. However, these treatments give rise to some degradation and removal of part of the outer primary wall of cotton fibre. In solution to damaging the cotton, also add to pollution load.

Keywords : Enzymes, Living organisms, Biological catalysts, De-grading colourants, Textile wet processing, Chemicals bracing.

7.1   Introduction

The synthetic dyes and chemicals are in use, since the time immemorial. However, today the use of some synthetic dyes and chemical is causing a serious threat to textile wet processing industry due to their carcinogenic nature. The recent ban on synthetic dyes by certain European countries has greatly affected the textile industries in India. Hence, systematic analysis and approach to control the pollution in this area has a very good expect on the eco-environmental condition.

7.1.1   Present state of art of the process

The Present state of Scouring involves hot alkali and other auxiliaries affect the rupture of the primary wall. Hot alkaline scouring results in environmentally harsh chemicals bracing added to the textile effluent. Bleaching with conventional (peroxide or chlorine) are effective in bleaching and degrading colourants and other impurities. However, these treatments give rise to some degradation and removal of part of the outer primary wall of cotton fibre. In solution to damaging the cotton, also add to pollution load.

7.1.2   Advance technology for scouring and bleaching using enzymes

Enzymes are nothing but biological catalysts. They are either a fungal or a bacterial source. Enzymes are present in living organisms but are themselves not living organisms. Structurally they resemble proteins. These proteins greatly accelerate reactions, but cannot cause a reaction to occur that ordinary would not occur, however, in the absence of enzymes an extremely long time may be required for the same reactions to occur. Initially, enzymes were used only for the removal of starch but nowadays enzymes are widely used in textile wet processing. Some researchers have carried out experiments with a lower concentration of cellulose in attempts to make the cotton fibre absorbent and succeeded in their attempt. However, the cellulose enzymes were not pure. Now it is revealed that pectinex enzyme is also mixed with cellulase, and which is responsible for the greatly improved water absorbency of cotton.

The pectin is powerful biological glue that binds the waxes and proteins together in the primary wall matrix and therefore hydrolysis of pectin in the primary wall matrix promotes an efficient interruption of the matrix to achieve superior water absorbency without the negative side effect of cellulose destruction. Enzyme treatments are safe to use because the only degrade one type of impurity and do not harm the fibre substrate. Pectinex, proteases degrade proteins, lipases degrade all type of waxes, and celluloses can be used to abrade the fibre surface by destroying cuticle and primary wall. Various research workers carry out experiments on the action of the above enzyme or in two or more combination on the impurities of cotton, which have the new way of scouring and bleaching using biological means.

7.1.3   Eco-pressure on synthetic chemicals and dyes

An up-gradation of textile processing technologies is essential to meet new challenges, an analysis of synthetic dyes and their dangers, banned chemicals, and alternatives for the future. The synthetic dyes and chemicals are in use, since the time immemorial. Perkin discovered the first synthetic dye a long time ago, but today the use of some synthetic dyes and chemicals is causing serious to textile processing industry due to their carcinogenic nature. The recent ban on synthetic dyes by Germany and certain European countries has greatly affected the textile industry in India. Eco-friendly dyes and chemicals are used worldwide to replace the carcinogenic amines. Therefore, ecological and environmental awareness is the immediate need to provide the people with pollution free environment.

7.1.4   German ban and eco-awareness

In Germany, it was found that symptoms of cancer occurred among workers who handled benzidine in the production of azo dyes. Aromatic amines on reduction give insoluble amine groups which on contact with skin and settles on the bladder. They affect the body metabolism and ultimately induce cancer. Due to this, several European dye-manufacturing companies stopped producing and marketing azo dyes based on benzidine in the 1970s. Later on, several tests were conducted on azo dyes and it was found that azo dyes could be split under the certain physiological condition to form carcinogenic amines. In July 1994, Germany amended two laws that were significant for textile processing industries. These were the Fourth Federal Emission Protection Ordinance, its aim being to protect the environment from the effluent of the textile industry and the ordinance of material and articles, its aim is to protect the consumer. This prohibits the use of certain azo dyes, which could split and give 20 aryl amines proven or suspected to be carcinogenic.

A red list and a green list has already been prepared in most of the European countries given listed items are labeled as eco-friendly. Red listed items are either been banned or being phased out. Today scientifically found evidence proves that four aromatic amines have a carcinogenic effect on the human. These are benzidine, 2-naphylamine, 4-aminobiphenyl, and 4-chloro-o-toluidine. These four amines are listed as definitely carcinogenic and placed in-group III A1 of German MAK list. The listed amines are also carcinogenic under the optimum condition of splitting and replaced in-group III A2 of MAK list. The list prepared by the German Committee for testing carcinogenic materials is called MAK list and is updated every year.

7.1.5   The Indian ban

The German ban on the import of textile and garments dyed with certain azo dyes initiated a number of activities in India by various organizations. Thus, many seminars have been held, technical articles published and against the ban on various grounds, including the one that test methods have not been prescribed by the German authorities and that maximum limits for the presence of the banned amines in the fabrics on extracting and reducing are not specified. Even it has been argued that the ban applies only to water-soluble dyes like direct, acid, basic, and other dyes and not to azoic pigments even though they are made from banned amines. A number of testing laboratories have been provided with large amounts of funds by the Indian Government to equip them with sophisticated instruments. Finally, the Ministry of Environment and Forests issued a notification on 29th March 1996. The notification stated that it applied in respect of the handling of azo dyes specified in the schedule of notification, the prohibition on the handling of azo dyes applied to the whole of India.

7.1.6   Replacement for carcinogenic dyes

For the development of non-carcinogenic dyes, the following points can be recommended:

1. The intermediate must be non-carcinogenic.
2. If in any case, carcinogenic intermediate is used, the carcinogenic intermediate should not be reformed from the dye by a reversible reaction during application or exposure to light.
3. To develop potentially new-chromophores giving the same colour and fastness properties as the azo dyes

7.1.7   The intermediate must be non-carcinogenic

The dyes based on carcinogenic amines benzidine, β-naphthylamine, and certain aromatic amines are banned, as these dyes on exposure to light or during the application, slowly break down to give the parent carcinogenic amine which can be harmful to end uses.

7.1.8   Need for ecofriendly processing

The textile industry is one of the major industries in India contributing nearly one-third of the countries total export earnings. The textile export from India is in the form of garments, fabric, and yarns used to produce the garments. The very interesting to note that garment exports bring the value addition to maximum extent. It is highly profitable to convert cotton into yarn and then to the garment and then export the same. This is the study increase in the textile export which is expected to reach around rupees 25,000 crores in this millennium. All these development coupled with the introduction of international eco standards, the introduction of ISO 9000 series certification process, drive against toxic substances and chemicals used in the fabric pose heavy responsibility on textile processing technologies is absolutely essential to meet new challenges in the international export market.

7.2   Enzymes the marvelous molecular machines

7.2.1   Historical developments

The term enzyme is derived from the Greek word enzymes, which means in the cell or ferments.' They are complex protein ferments secreted by living organisms. Enzymes are the vital parts of all the living processes and are believed to as old as life itself. Enzymes are specialized biopolymers composed of many different amino acids that have complex three-dimensional structures held in place by varieties of bonding forces. They are naturally occurring high molecular weight proteins capable of catalyzing the chemical reactions of biological process and hence known as biocatalysts. Enzymes are natural organic molecules consisting of three-dimensional proteins made up of alpha amino acids. The three-dimensional shape gives them the capability to behave as catalysts in increasing the rates of specific chemical reactions.

Enzymes can replace harsh chemical processes used in the textile industry and may catalyze reactions at ambient temperature reducing the need for high temperature and pressure otherwise required. From a chemical standpoint, an enzyme is defined as a protein complex composed of about 200 to 250 amino acids. The molecular weight of these ferments is very high and is of the order of 10^4-10^5. They are specific biochemical catalysts made to take part in a particular reaction, being redecorated in their original form at the end of a particular reaction. The major difference of enzymes from the chemical catalysts is that they are temperature sensitive, having relatively low energy of activation and are usually active over a narrow range of pH. The Enzyme reacts only with a specific substrate, which fits within the active site of the enzyme molecule. Enzymes are relatively fragile substances with a tendency to undergo de-nurturing and inactivation under unsuitable condition.

The application of enzymes in the textile industry is becoming increasingly popular because of their use under mild conditions of temperature, pH, and their capability of replacing harsh organic chemicals. Typical temperatures of processing during enzymatic treatment are about 40-50°C, which confers a significant decrease in energy consumption compared with normal processing temperatures. Also important is that wastewater from enzymatic treatments is readily biodegradable and accordingly does not pose any environmental threat. The most typical applications of enzymes for treatments of cotton fibres can be summarized as follows:

- Desizing: Removal of starch with amylases.
- Souring: Dissolution/dispersion of waxes, proteins, pectins, and natural fats from the surface of the cotton fibres with amylases- lipase cellulase solution.
- Bleach cleaning:Removal of residual hydrogen peroxide with catalase.
- Bio-polishing: Improvement of the appearance of the cotton fabric by removal of fuzz- Fibres and pills from the surface with cellulose.

7.2.2 Mechanism of enzyme action

The action of the enzyme, first of all, forms an enzyme-substrate complex. Direct physical contact of the enzyme and substrate is required to obtain the complex. However, the mechanism of enzymatic hydrolysis of cellulosic materials is complicated and not yet fully understood. Enzymes contain a true activity centre in the form of three-dimensional structure like fissures, hole, pockets, cavities, or hollows. The number of active sites per molecules is very small, generally only one. The overall rate of reaction depends on the times required to form the enzyme-substrate complex and the time required to form the final product. In order to catalyse a reaction, the enzyme molecule has to form a complex with the substrate.

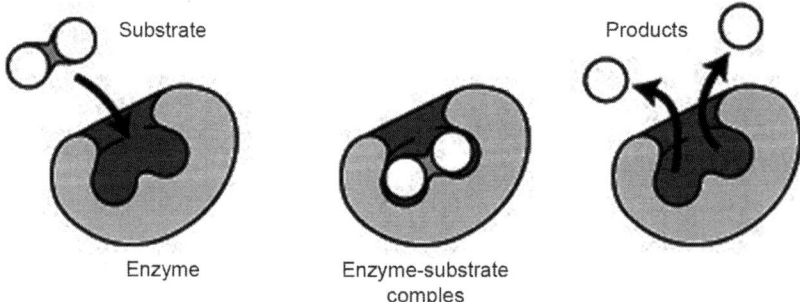

<table>
<tr><td>Enzyme</td><td>Enzyme-substrate comples</td><td></td></tr>
</table>

To form the complex product, the binding and reactive sites of enzyme recognize the corresponding domain of the substrate molecules. This makes the enzyme specific towards the substrates. Exceedingly high substrate concentration has an adverse effect on the enzyme reaction. This substrate crowding creates a bottleneck at binding sites or reactive sites or both resulting in lowered reaction rates. Finally, the complex disintegrates with the release of the reaction products and the original enzymes are once again available. Synergism of different components in the enzyme complex and inhibition mechanisms further complicates the reaction. Enzyme diffusion plays a much more decisive role in the heterogeneous system of soluble enzyme solid substrate. The kinetics of the reaction, therefore, depends on the diffusion of the enzyme into the solid phase of the substrate and the diffusion of the reaction products out of their solid phase into liquor. For cotton, the restriction of the enzyme to the fibre surface is easily achieved, because cellulose is a highly crystalline material and possesses only small amorphous areas, making the diffusion of enzymes into the interior of the fibre nearly impossible. Thus, by regulating enzyme dosage and choosing the type of enzyme, it can be confined to the surface of cotton and to the amorphous regions, leaving the fibres intact.

7.2.3   Kinetics of enzyme action

Enzymatic hydrolysis follows the kinetics of first-order equation. The kinetics measured during a textile process can provide useful mechanistic information. An international enzymatic unit called KATAL is defined as the quantity of enzyme, which transforms one mole of substrate per second under standard conditions. Submultiples more commonly used are the microkatal (μ Kat), the nanokatal (n Kat) and the picokatal (p Kat) expressed respectively as 10^{-6}, 10^{-9} and 10^{-12} katal (Kat). This unit is seldom used in practice. The kinetics of enzymatic catalysis follows the Michaels-Mention equation that in it's simplified from for a reaction is written as follows;

$$E + S = (E - S) - E + P$$

The substrate S forms a transitory intermediate complex with the enzyme E, that is, (E – S). By applying the classical Michaels-Mention equation, one can calculate the rate of substrate turnover at saturation (V_{max}) and a half-saturation constant (K_m), which can be interpreted as an apparent dissociation constant of all enzyme bound species. However, expressing the reaction rate (V) as a function of substrate concentration(S) is a very artificial convention when the reaction takes place on hydrated cellulose, where it is impossible to change the concentration of the substrate sites. Moreover, it is verified that the EG-CBH synergism disappears at high enzyme concentration, that is, it is competitive synergism. For these reasons, it is suggesting that (V) should be expressed as a function of enzyme concentration (E).it has been suggested that an analogous equation should be used. The calculated parameters Vmax and Ke are analogous to the Michaels-Mention parameters.

$$V = (V_{max} [S] / (K_m + [S])$$
$$V = (V_{max} [E] / (K_e + [E])$$

7.2.4　　Enzymatic activity

The enzymes currently used in the textile industry, namely α-amylases, cellulases, and proteases are included in the class of hydrolases. These enzymes provoke the hydrolysis of various types of linkages such as ester, acetyl, amide, etc…The enzyme splits various compounds and becomes fixed on the residues of the elements in the water. The efficiency of the enzyme-catalyzed reaction depends on the following factors:

7.2.5　　Temperature

Enzymes are observed to exhibit maximum activity in a narrow range of temperature. The effect of temperature on the enzyme activity is very complex and interlaced with other variables such as pH, buffer system and substrate concentration. It is hence imperative that a thermo-stable enzyme must be used if the reaction temperature is not high, and then less expensive enzymes, which are not thermo-stable, could be conveniently selected.

7.2.5　　Presence of hydrogen (pH)

Most enzymatic activities are extremely sensitive to the pH and its variations. Enzymes being composed of proteins, exhibit zwitterions properties. The proton donating or proton accepting groups in enzyme catalytic sites are at their required state ionization at a selected pH of the enzymes at which its activity is optimal. Any increase or decrease of pH results in lowered reaction rates. So, the representation of the speed of an enzymatic reaction as a function of the pH is normally a bell-like curve comprising of a phase of incipient enzymatic activity, one of maximum activity and one of declining activity, with the mid-point called to the optimum pH.

A variation of pH during the course of reaction may bring about an alteration in the protein structure with a denaturing effect on enzymatic or the ionization of the active site. The optimum pH also varies from one enzyme to another. In order to achieve a high yield from the enzymatic catalyst, it is important to adjust the pH to an optimum value in each case and buffer the medium to void deviations of the pH during the course of the reaction.

7.2.6　　Activators

Certain enzymes require metal ions as activators, generally bivalent metallic cations. These metals stabilize the structure of the enzymes substrate complex or sensitize the substrate to the attack of the enzyme or again make the action of the ferment more efficacious. Certain metals are capable of becoming part of the enzyme and they enter into the constitution of the prosthetic group and take part in the ion exchanges. One can even consider the activator ion as the prosthetic group itself. This is the case with the hydrolysis of starch by a-amylase, in the presence of chloride ion and the cationic Ca^{2+}. Certain wetting agents also exhibit stabilizing action on enzymes. Only those commercial products, which do not disturb the pH of the system, should be selected. Thus, non-ionic agents are generally preferred.

7.2.7   Inhibitors

Certain chemicals such as alkalis, antiseptics, and acid liberating salts tend to inhibit the enzyme activity, which could be either reversible or irreversible. These inhibitors to enzyme activity act in different ways, for example, by blocking certain useful groups. The reaction products themselves may inhibit the enzyme. With the competitive type of inhibition, the inhibited processes an affinity for the enzyme with which it is combined, thus actually creating competition between itself and substrate. Its action is reversible. In the non-competitive type of inhibition, the inhibition substance associates itself with the ferment without being disturbed and without being competed with the enzyme substrate combination. These agents reduce V_{max} of the enzymatic catalysts. In general, the cations of heavy metals of lethal to enzymes and their effects detrimental. Few examples are salts of lead, mercury, copper, iron, etc. reducing and oxidizing agents are also known to act as inhibitors of enzymes, as their oxidizing – reducing properties can destroy the enzyme molecule. Sequestering agents are also lethal to catalase or peroxydase types.

7.2.8   Incubation period

The duration of reaction is of tremendous importance as it enables to understand whether hydrolysis or oxidation in the substrate is carried out completely. Therefore, for the development of a successful commercial process, care must be exercised without overlooking the factors influencing the activity of enzymes.

7.3   Enzyme treatment on cellulosic fibres

Commercial enzyme treatments have generally targeted cotton. There is no explicit limitation to cotton; however, little has been reported on other cellulosic fibres. Although the chemical composition is identical, there are major differences in the fine structure and morphology of these fibres, which actually determines the course of the enzymatic degradation reaction. Linen and ramie have higher Crystallinity and the pitch of the spiral structure is less than cotton. Furthermore, spiral reversal and convolutions occur in cotton. Significant differences in their pore structure and crystallite sizes have also been found. Bast fibres like linen and ramie are multiple cellular systems, in contrast to cotton, which consists only of a single cell. Multi-cellular fibres contain natural gums and resins that keep the cells together. It can influence the course of enzymatic hydrolysis. Regenerated cellulosics, on the other hand, are much simpler systems than natural cellulosic fibres. The viscose rayon has considerably lower DP, Crystallinity, and orientation than cotton.

Enzymatic hydrolysis can be used to decrease stiffness, ease stretchability and to loosen the structure of cellulosic fabrics. Cotton, linen and ramie fabrics manifest the desired effect of removal of surface fibrils without suffering large weight losses or reductions in tensile strength by enzymatic treatment. The strength/ weight loss relations of ramie and linen differ from those of cotton and viscose rayon. It is seen that the crystalline index of the samples does not change after the enzymatic hydrolysis. This suggests that the ratio of crystalline to the less ordered region does not change upon enzymatic degradation.

The cellulose, xylanase, and pectinase enzymes have a tremendous effect on the processing of jute. The treatment with enzymes before bleaching of jute improves whiteness, whereas due to back staining at optimum pH, there is a decrease in whiteness and increase in yellowness index if the treatment is carried out after bleaching. The enzyme reaction is also seen more on 4% sodium hydroxide scoured fabric. Scouring causes loss of hemicellulose, producing an open structure and thus a larger surface area of lignin is accessible to H_2O_2, resulting in higher whiteness.

7.4   Structural changes of fibres

The cellulose complex diffuses through the pore system to the microfibrils, attacks the cellulose chains, and hydrolyses each chain to the end. The differences in the efficiency of cellulases on various fibres are dependent on a number of factors such as the amount of non-cellulosic matter, the degree of polymerization type and the number of chemical substitutions to the cellulose. Key features of the cellulose substrate are Crystallinity, accessible surface area and pore dimensions.

7.4.1   Advantages of enzyme treatment

Recently, enzymes are widely being used in textile applications. The use of enzyme in degrading starch has been well known. Enzyme technology is great interest in the chemical demanding pretreatment of cotton, wool and silk fibres. Cellulases can be used for finishing treatments on cotton, leading to fabric softness, good performance properties, and fashionable looks. It also has the potential to simplify and cheapen manufacturing processes in an environmentally friendly way. The following benefits are achieved in practice:

- The enzymatic process removes the small fibre ends found in the yarn surface, which eventually leads to pilling on the fabric surface.
- The fabrics treated with cellulases are free from the surface hariness , neps, and fluff with much improved handle and flexibility.
- The material sticking is prevented particularly with mercerized fabric along with the improvement in texture and sewability.
- The effect of enzymatic treatment is long-lasting. The colour of the dyed goods becomes brighter with a visually improved colour yield.
- Cellulases have been incorporated in detergents to remove the fibre fuzz.
- Another advantage of enzymatic processes is that they can be adopted to run equipment already existing in textile processing units.

7.4.2   Drawbacks and remedie

The disadvantages still unsolved in the practical application of the cellulose treatment is that the cellulose catalytic reaction rate is affected appreciably not only by pH and temperature but also by existing chemicals such as dyes or surfactants in the treatment solution or on the substrate. Another problem with enzymes is the wastage during processing in the form of a residual bath, as the enzymatic action on cotton is a slow and time-consuming process. Standing bath technique can successfully be used to reuse the residual enzyme bath. Pretreatment of cotton with various swelling agents can be used as a tool to enhance the response of enzyme towards cellulosic substrates. The process ultimately reduces the amount and thus the cost of the enzyme to be applied for achieving the desired effect.

7.4.3   Enzymatic desizing

Amylase enzymes are commercially available with flexible pH and temperature ranges, with some products even being thermostable. A whole amylase complex usually contains various types of exo and endo-enzymes, glucoamylases and debranching enzymes with different modes of action. Starches cannot be recycled. The end products of the enzymes desizing process are various types of sugars and dextrin's which are non-toxic, however, negatively impact the BOD of the wastewater.

The desizing process can be divided into three stages: Impregnation, Incubation, and Afterwash. The fabric may be pre-washed to remove non-starch water-soluble additives and to facilitate the binding of amylase to the starch molecules, thorough wetting, and heating to gelatinize the starch facilitates the contact between the enzymes and its substrate. Impregnation should be carried out at a temperature above 70°C in buffered solution calcium. Alternatively, the fabric may be soaked with enzymes solution at the optimum temperature before a longer incubation is carried out. The incubation stage may take 2-16 hours, depending upon the stability and the activity of the enzyme at the processing temperature and pH, the nature of the size and type of fabric.

The efficiency of size removal is determined using the iodine colour test to measure residual starch. A dilute solution of iodine in water gives a deep blue-black colour presence of starch. If starch is absent, the solution reveals pale yellow-brown colour. The test is very sensitive; traces of starch are readily detected. Desizing may be improved with a combination of amylase and lipase treatment. A mixed enzyme treatment was found to: improve the removal of starch and triacylglycerol, give less risk of dye defects and a higher whiteness after scouring and bleaching, give a higher uptake of dye, to have a softening effect on the fabric.

7.4.4   Bio-scouring

An innovation approach to the problem of removing fibre impurities from cotton textile material is now emerging from great advances that are made in biotechnology. The majority of cellulose of cotton fibre is present in the secondary wall, which has an internal canal termed as the lumen. It contains the remains of protein impurities present during the fibre growth period. The outer layer primary wall and cuticle consists of proteins, glucans, and pectinex substances, along with a wax complex of fatty acids and high molecular weight alcohols and esters. These compounds are responsible for the poor absorbency of cotton. They account for more than 50% in the primary wall composition. In any type of scouring, the main aim is to rupture the primary wall and cuticle in order to remove them from the fibre. Traditional preparation involves hot alkali and other auxiliaries to affect the rupture of the primary wall. If necessary precautions are taken oxycellulose formation happens resulting in fibre strength loss. In addition, hot alkaline scouring results in environmentally harsh chemicals being added to the textile effluent.

Enzymes are nothing but biological catalysts. They are either a fungal or a bacterial source. Enzymes are present in living organisms but are themselves not living organisms. Structurally they resemble proteins. These proteins greatly accelerate reactions, but cannot cause a reaction to occur that ordinary would not occur, however, in the absence of enzymes an extremely long time may be required for the same reactions to occur. Initially, enzymes were used only for the removal of starch but nowadays enzymes are widely used in textile wet processing. Some researchers have carried out experiments with a lower concentration of cellulose in attempts to make the cotton fibre absorbent and succeeded in their attempt. However, the cellulose enzymes were not pure. Now it is revealed that pectinex enzyme is also mixed with cellulase, and which is responsible for the greatly improved water absorbency of cotton.

The pectin is powerful biological glue that binds the waxes and proteins together in the primary wall matrix and therefore hydrolysis of pectin in the primary wall matrix promotes an efficient interruption of the matrix to achieve superior water absorbency without the negative side effect of cellulose destruction. Enzyme treatments are safe to use because the only degrade one type of impurity and do not harm the fibre substrate. Pectinex, proteases degrade proteins, lipases degrade all type of waxes, and celluloses can be used to abrade the fibre surface by destroying cuticle and primary wall. Various research workers carry out experiments on the action of the above enzyme or in two or more combination on the impurities of cotton, which have the new way of scouring and bleaching using biological means.

Scouring of cotton could be carried out by the combination of pectinases and cellulases . They have tested the enzyme treated samples for staining test, microscopic observation, and the absorbency tests. It was concluded that pectinases penetrate the cuticle in aqueous solutions through cracks or micro-pores and hydrolyze the pectic substances. Thus, cuticle is practically removed and the continuing of the cuticle is broken down. The wax and protein like impurities are discarded by the fibre and emulsified. Cellulases penetrate the cuticle in aqueous solutions through cracks or micro-pores in the cuticle and make contacts with the primary wall. It catalyses hydrolyses of part of the primary wall at the contact point. The cellulose can catalyze the hydrolysis of a more amorphous primary wall than the crystalline secondary wall because natural crystalline cellulose is very resistant to cellulose digestion. The primary wall is partly hydrolysed and therefore a break of the linkage between the cuticle and the cellulose body is broken, which results in the loosening of all the impurities.

A mixture of cellulases and pectinases in a scouring bath act separately as above. Cellulases break the linkage from the cellulose side and the pectinases break the linkages from cuticle side. The action of pectinases creates more sites in the primary wall layer available for cellulose digestion and the action of cellulases creates more sites in the pectic substance layer available for pectinase digestion. This results in a more effective scouring in terms of speed and evenness. They have also studied the effect of lipases and proteases but found them ineffective compared to pectinases and cellulases. Alkaline pectinase is more useful in scouring, has a three-dimensional structure and is absorbed onto the three-dimensional pectin substrate in lock-and-key fashion. Hydrolysis of pectin is very strongly accelerated at the interface between the pectinase/pectin complexes. Once the hydrolysis in over at that site, pectinase enzyme is released and is attached to another area of the pectin substrate. The process of hydrolysis is repeated in different regions of the primary wall, which results in a cotton fibre absorbent. After enzymatic treatment, requisite quantities of water and buffer solution were added to maintain the same concentration and pH. Thus, a standing bath is maintained. The treated samples are tested for the weight loss and sinking time (absorbency) compared to the fresh bath, in the replenishment bath weight loss decreases which means that the efficiency decreases in every subsequent bath. The extent of the decrease in efficiency is very less(less than 1%) in every bath and therefore negligible. This technique of reuse of standing baths will reduce the cost.

7.4.5   Bio bleaching

Oxidative chemical pre-treatments are very effective in bleaching and degrading colourants and other impurities. However, such pre-treatments always give rise to some oxidative degradation and removal of part of the outer primary wall of cotton fibre. Some of the chlorine-based oxidative bleaching agents in addition to damaging the cotton also add to pollution load. This the bleacher always aims at maximum whiteness, wettability, and absorbency with minimum fibre damage. Bleaching possibilities of using peroxidases lacase/mediator systems and glucose oxidants are explored. Physical and the fine structural properties of textiles after these treatments are investigated. Laccase/mediator systems have been used as enzymatic bleaching. The enzymes can activate various oxidizing agents such as hydrogen peroxide. These enzymes are deactivated rapidly during the bleaching process and therefore a fully satisfactory bleaching effect has not been established. The third and most promising approach is the use of glucose oxidases. These enzymes generate hydrogen peroxide and glycolic acid from glucose and oxygen. This type of enzymes seen the most suitable of all the three use the sugar contaminated effluents from the other wet processing steps.

7.4.6   Bio polishing

The newest method of handle modification for cellulosic fibres is treated with specific enzymes. As this is a biological process names such as "bio-finish" and "bio-polishing" are used for this process. Cellulases can catalase the hydrolytic splitting of the glucoside bond in cellulose. A cellulose preparation is produced from non-pathogenic fungi. These preparations contain exo-cellulase, endo-cellulase, and β-glucoside, which give the enzymes the ability to modify the cellulosic fibre. The aim of bio finishing is to shorten the molecular chain, which leads primarily to a loss in tensile strength. However, such enzymes are large molecular complexes they cannot penetrate the inside of the textile material. They act mainly on the surface and remove free fibre ends (hairiness).under the influence of mechanical treatment these fibres break off and produce a smooth surface of the fabric.

This treatment also loosens the material structure leading to a slightly softer handle. In most cases, the difference can be distinguished, only after giving a final regular cationic finish. However, one should not expect too much from bio-finish. The assertion that seconds due to hairiness can be made into first-class goods by this process clearly goes beyond the realities of this process. Bio finishing is normally carried out in a batch process since control is very easy. It is carried out either before dyeing (after bleaching) or at the final stage.

Advantages of Bio-polishing:
- Reduced hairiness and fluff on fabric.
- Improved softness and drape.
- Increased hydrophilic properties.
- Improve elasticity of fabric
- Increase dye affinity, yield, and levelness

7.5   Materials and methods

7.5.1   Processing of cotton woven fabric

Cotton fabric has several impurities such as fats, waxes, pectinous substances, and ash, Presence of fats and waxes in cotton fabric imparts poor absorbency .in order to remove these materials many processes are carried out. The various processes are desizing, scouring, bleaching, dyeing, and finishing are discussed below.

Normal process:

For normal chemical process, materials and process parameters involved are given below.

Material	:	Woven fabric
Weight of material	:	120 kgs
Quantity of wate	:	250 liters
Machine	:	Jigger
M : L	:	1:2

Desizing:

Desizing is the process of removing the sizing materials. Sizes are applied to withstand stress and strain during weaving. In the conventional process, the oxidizing agents are used for desizing. The various chemicals and parameters involved in desizing are given below.

Tinozyme L 40	:	0.75%
Wetting agent	:	0.30%
Temperature	:	80°C
Duration	:	2 ends (1 end= 15 mins; 2 ends= 30mins)

Combined scouring & bleaching:

In normal process, the scouring of cotton fabric is normally carried out with strong alkali at high temperature and for longer duration. The various chemicals and parameters involved in scouring and bleaching are given below.

Caustic	:	2.0%
H_2O_2	:	4.0%
Soda	:	1.0%
Silicate	:	1.0%
Wetting agent	:	0.5%
Temperature	:	100°C
Duration	:	4 ends (1 hrs)

Enzyme process:

The Textile industry has a wide potential for adaption of enzyme technology in its processes like desizing, bio-washing, biopolishing, and effluent treatment. For biochemical process, materials and process parameters involved are given below.

Material	:	Woven fabric
Weight of material	:	120 kgs
Quantity of water	:	250 liters
Machine	:	Jigger
M : L	:	1:2

Desizing:

Desizing is the process of removing the sizing materials. Sizes are applied to withstand stress and strain during weaving. The various chemicals and parameters involved in desizing are given below.

Tinozyme L 40	:	0.75%
Wetting agent	:	0.30%
Temperature	:	80°C
Duration	:	2 ends (1 end= 15 mins; 2 ends= 30mins)

Combined scouring & bleaching:

The bio-preparation of scouring is the removal of oils, fats, waxes, and pectic substances and to improve the dyeing characteristics. The various chemicals and parameters involved in scouring and bleaching are given below.

Bio-scour GX-1	:	2%
H_2O_2	:	4%
Caustic	:	0.1%
Temperature	:	100°C
Duration	:	2 ends

Dyeing:

The dyeing processes for cotton, a preferred fibre for apparel are not especially environmentally friendly. The typical reactive dye has only low to moderate affinity for the fibres and exhaust procedures with these dyes require electrolytes to enhance dye-fibre interactions and lengthy after-washing to achieving the wash-fastness properties. The dyes and chemicals used for dyeing is given below.

Levafix yellow CA (Reactive dye)	:	0.025%
Levafix red CA (Reactive dye)	:	0.12%
Salt	:	10gpl
Soda	:	5 gpl

Softening:

The aim of softening is to shorten the molecular chain. The enzymes and chemicals are applied to cellulosic fabrics and garments to remove the surface protruding fibres to achieve soft-smooth feel, lustrous, fuzz-free surface, and anti-pilling property.

Normal chemical Softening:

The various chemicals and parameters involved in normal softening are given below.

PISOFT PC 100	:	30 gpl
Quantity of water	:	10 liters
Temperature	:	Room temperature
Duration	:	30 mins
M:L	:	1:2

Bio Softening:

The various chemicals and parameters involved in bio softening are given below.

BACTOSOL CA	:	0.3%
Quantity of water	:	10 liters
Temperature	:	50°C +/– 5°C
Duration	:	30 mins
M : L	:	1:2

Processing of Cotton knitted fabric:

Cotton fabric has several impurities such as fats, waxes, pectinous substances, and ash, Presence of fats and waxes in cotton fabric imparts poor absorbency .in order to remove these materials many processes are carried out. The various processes are scouring, bleaching, and dyeing are discussed below.

Normal process:

For normal chemical process, materials and process parameters involved are given below.

Material	:	Knitted fabric
Machine	:	winch
M : L	:	1:10

Combined Scouring & Bleaching:

The scouring process is carried for to remove the oils, fats, waxes, and pectic substances and to improve the dyeing characteristics. In the normal process, the scouring of cotton fabric is normally carried out with strong alkali at high temperature and for a longer duration. The various chemicals and parameters involved in scouring and bleaching are given below.

Caustic	:	2.0%
H_2O_2	:	6.0%
Soda	:	1.5%
Silicate	:	1.5%
Wetting agent	:	0.5%

Temperature	:	100°C
Duration	:	1 hr

Enzymatic process:

The Textile industry has a wide potential for adaption of enzyme technology in its processes like bio-scouring, bio-washing, bio-polishing, and effluent treatment. For the biochemical process, materials and process parameters involved are given below.

Material	:	Knitted fabric
Machine	:	winch
M : L	:	1:10

Combined Scouring & Bleaching:

The bio-preparation of scouring is the removal of oils, fats, waxes, and pectic substances and to improve the dyeing characteristics. The various chemicals and parameters involved in scouring and bleaching are given below.

Bio-scour GX-1	:	2.0%
H_2O_2	:	6.0%
Alkali	:	0.2%
Temperature	:	100°C
Time	:	1 hr

Dyeing:

The dyeing processes for cotton, a preferred fibre for apparel are not especially environmentally friendly. The typical reactive dye has only low to moderate affinity for the fibres and exhaust procedures with these dyes require electrolytes to enhance dye-fibre interactions and lengthy after-washing to achieving. the wash-fastness properties. The dyes and chemicals used for dyeing is given below.

Levafix yellow CA (Reactive dye)	:	0.025%
Levafix red CA (Reactive dye)	:	0.30%
Salt	:	10 gpl
Soda	:	5 gpl

7.4   Methodology

Methodology flow chart of the work for existing and enzymatic methods:

The flow chart shows the steps involved in the conventional chemical processing of cotton woven fabric.

Normal process sequence (woven fabric):

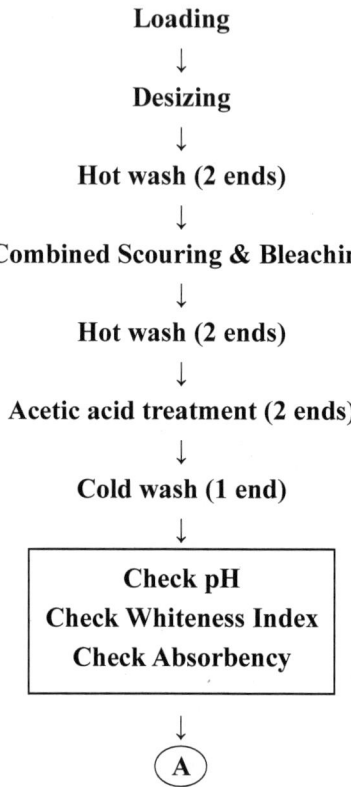

Loading
↓
Desizing
↓
Hot wash (2 ends)
↓
Combined Scouring & Bleaching
↓
Hot wash (2 ends)
↓
Acetic acid treatment (2 ends)
↓
Cold wash (1 end)
↓

> Check pH
> Check Whiteness Index
> Check Absorbency

↓
(A)

Enzymatic process sequence (woven fabric):

The enzymatic process sequence is shown below. This process is carried out in a jigger machine.

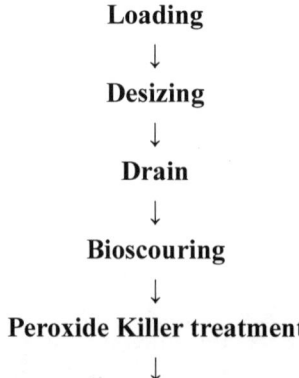

Loading
↓
Desizing
↓
Drain
↓
Bioscouring
↓
Peroxide Killer treatment
↓

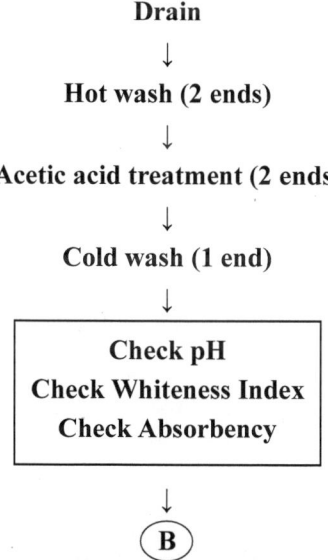

Drain

↓

Hot wash (2 ends)

↓

Acetic acid treatment (2 ends)

↓

Cold wash (1 end)

↓

Check pH
Check Whiteness Index
Check Absorbency

↓

(**B**)

The normal process sample (**A**) and the enzyme process sample (**B**) are attached in the same dyeing bath. The dyeing process is carried out. Here pink shade is taken for trails by using reactive dyes. Dyes and chemicals are weighed accurately dissolved properly.

½ Qty of dyes added. Run the jigger for 1 end

↓

Add balance ½ Qty of dyes. Run the jigger for 1 end

↓

½ Qty of salts are added. Run the jigger for 1 end.

↓

½ Qty of salts are added. Run the jigger for 1 end.

↓

Temperature is raised to 60 C, run the jigger for 2 ends.

↓

½ Qty of soda is added. Run the jigger for 1 end.

↓

Add balance ½ Qty of soda. Run the jigger for 1 end.

↓

Fixation is carried out at 4 ends at 60°C.

↓

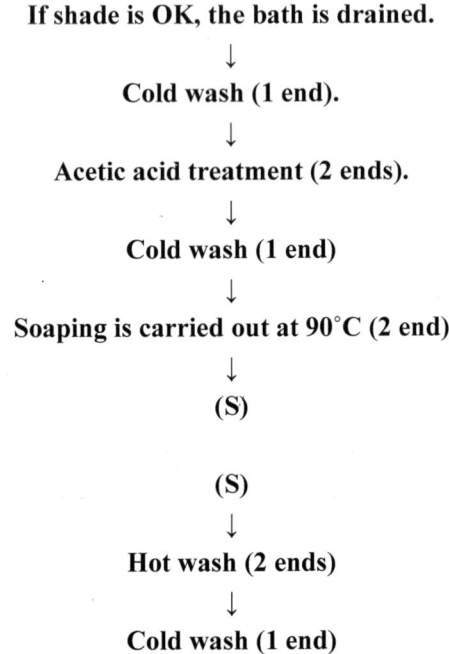

If shade is OK, the bath is drained.

↓

Cold wash (1 end).

↓

Acetic acid treatment (2 ends).

↓

Cold wash (1 end)

↓

Soaping is carried out at 90°C (2 end)

↓

(S)

(S)

↓

Hot wash (2 ends)

↓

Cold wash (1 end)

Then the normal process sample (A) and enzymatic process sample (B) are separated.

Softening:

The aim of softening is to shorten the molecular chain, which leads primarily to a loss tensile strength. The softening process is shown below..,

- The normal process sample (A) is softened by using normal cationic softener (<u>PISOFT PC 100</u>)
- The enzyme treated sample (B) is treated with <u>BACTOSOL CA</u> for softening purpose.

Then the sample (A) & (B) are taken out and dried.

Normal process sequence (knitted fabrics):

The flow chart shows the steps involved in conventional chemical processing of cotton woven fabric.

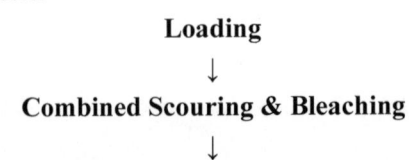

Loading

↓

Combined Scouring & Bleaching

↓

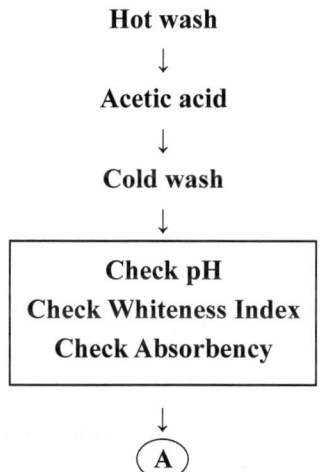

Hot wash
↓
Acetic acid
↓
Cold wash
↓

> **Check pH**
> **Check Whiteness Index**
> **Check Absorbency**

↓
(A)

Enzymatic process sequence (knitted fabrics):

The enzymatic process sequence is shown below. This process is carried out in a winch machine.

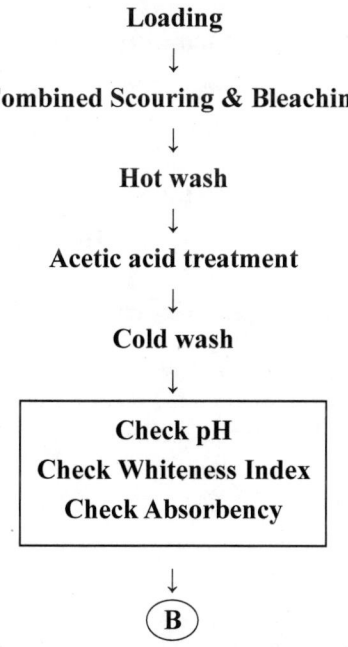

Loading
↓
Combined Scouring & Bleaching
↓
Hot wash
↓
Acetic acid treatment
↓
Cold wash
↓

> **Check pH**
> **Check Whiteness Index**
> **Check Absorbency**

↓
(B)

The normal process sample (A) and enzymatic process sample (B) are dyed in the same dye bath.

Dyeing

The normal process sample (A) and enzymatic process sample (B) are dyed in the same dye bath.

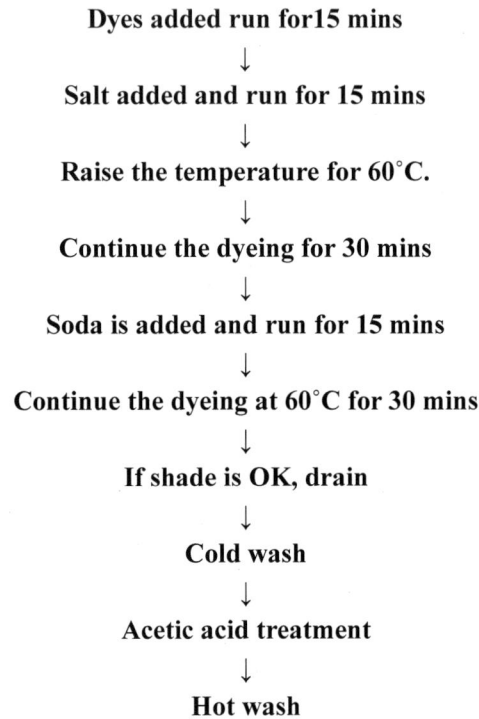

Dyes added run for15 mins
↓
Salt added and run for 15 mins
↓
Raise the temperature for 60˚C.
↓
Continue the dyeing for 30 mins
↓
Soda is added and run for 15 mins
↓
Continue the dyeing at 60˚C for 30 mins
↓
If shade is OK, drain
↓
Cold wash
↓
Acetic acid treatment
↓
Hot wash

The trial samples are substituted for testing.

7.5   Results and discussions

The normal scouring and bleaching methods are compared with the enzymatic method by preparing samples and their characteristics were tested using different fabric tests. The enzymatic method reduces toxic and weight loss in scouring and bleaching. It is also noticed that there is an increase in strength, whiteness, and absorbency properties of the samples.

The following tests have been carried out to assess the samples processed from normal and enzymatic methods.

- Wash fastness & staining fastness test.
- Absorbency test
- Bursting strength test
- Tensile strength test

- Dye strength comparison
- Whiteness Index

7.5.1   Test results

The normal scouring and bleaching methods are compared with the enzymatic method by preparing samples and their characteristics were tested using different fabric tests. The test results have been discussed below.

7.5.1.1   Wash fastness test& staining fastness test

The ability of fabric of not being faded even after frequent washes. The dyed fabric is attached to the undyed fabric of the same material. Then the specimen is placed in washing m/c and washing of the sample is carried out. The test results are shown in the table.

Cotton woven fabric

This table 7.1 (a) gives wash fastness grade for normal and enzyme-treated woven fabric.

Table 7.1 (a): Wash fastness grade for normal and enzyme-treated woven fabric.

Sample no	Normal processed fabric	Enzyme processed fabric
1	3.5	4
2	3.2	3.9
3	3.4	4
4	3.5	4.2
5	3.6	4.2
AVG	3.4	4

Cotton knitted fabric:

This table 7.1 (b) gives wash fastness grade for normal and enzyme-treated knitted fabric.

Table 7.1 (b): Wash fastness grade for normal and enzyme-treated knitted fabric.

Sample no	Normal processed fabric	Enzyme processed fabric
1	3	4
2	3.2	4.1
3	3.1	4
4	3	4
5	3.3	4.2
AVG	3.1	4

- In the case of woven and knitted fabrics, the enzyme-treated fabric has slightly high wash fastness properties

7.5.1.2   Staining fastness test

The dyed fabric is attached to the undyed fabric of the same material. Then the specimen is placed in washing m/c and washing of the sample is carried with following as for the recipe. The test results are shown in the table.

For Cotton woven fabric:

This table 7.2 (a) gives staining fastness grade for normal and enzyme-treated woven fabric.

Table 7.2 (a)

Sample no	Normal processed fabric	Enzyme processed fabric
1	4.3	4.5
2	4.2	4.4
3	4	4.2
4	4.1	4.3
5	4.4	4.5
AVG	4.2	4.4

For Cotton knitted fabric:

This table 7.2 (b) gives staining fastness grade for normal and enzyme-treated knitted fabric.

Table 7.2 (b)

Sample no	Normal processed fabric	Enzyme processed fabric
1	3	3.5
2	3.2	3.9
3	3.4	4
4	3	3.4
5	3.1	3.6
AVG	3.1	3.7

- In case of woven and knitted fabrics, the enzyme treated fabric has slightly high staining fastness properties,

7.5.1.3   Absorbency test (sink test):

This test is carried out for the scoured fabric. The time taken of fabric to completely dip in the water with using of suitable beaker contains the required amount of water.

Cotton woven fabric:

This table 7.3 (a) gives absorbency characteristics of normal and enzyme treated woven fabric.

Table 7.3 (a)

Sample no	Normal processed fabric(sec)	Enzyme processed fabric (sec)
1	7	5
2	6	4
3	8	5
4	7	6
5	8	4
AVG	7.2	4.8

Cotton knitted fabric:

This table 7.3 (b) gives absorbency characteristics of normal and enzyme treated knitted fabric.

Table 7.3 (b)

Sample no	Normal processed fabric(sec)	Enzyme processed fabric(sec)
1	4	2
2	3	3
3	5	2
4	4	3
5	5	4
AVG	4.2	2.8

- The absorbency of enzyme treated fabric is high, because the enzymes will not attack the cellulose substrate.

7.5.1.4   Bursting strength test

Bursting strength is the strength of fabric against a multi-directional flow of pressure. The bursting test measures a composite strength of both warp and weft yarns simultaneously and indicates the extent to which a fabric can withstand. The reason for this method of testing may be due to the material in use is stressed in many directions simultaneously. The pressure in a liquid is exerted in all directions and advantage is taken of this phenomenon in the hydraulic bursting tester. This table 7.4 gives bursting strength characteristics of normal and enzyme-treated knitted fabrics.

Table 7.4

Sample no	Normal processed fabric(lbs/sq.inch)	Enzyme processed fabric(lbs/sq.inch)
1	50	95
2	55	95
3	50	100
4	60	100
5	65	95
6	50	100
7	55	95
8	60	90
9	50	100
10	55	95
AVG	50	87.73

- The enzyme-treated fabric bursting strength is double when compared to the normal processed fabric.

7.5.1.5   Tensile strength test

The breaking strength is a measure of the resistance of the fabric to a tensile load or stress in either warp or weft direction. This test is carried out for cotton woven fabric. This indicates that the strength of fibres and fibre structures is commonly regarded as the criterion of quality. The test results are shown in the table.

Cotton Woven fabric

This table 7.5 gives tensile characteristics of normal and enzyme treated woven fabric.

Table - 7.5:

Sample No	Direction	Normal processed fabric(lbs)	Enzymatic processed fabric(lbs)
1		35	45
2		32	44
3		34	43
4		35	46
5	Warp	33	44
6		18	23
7		15	26
8		19	25
9		17	25
10	Weft	18	24

Tensile strength of normal processed fabric in

> Warp way : 34 lbs
>
> Weft way : 17 lbs

Tensile strength of enzymatic processed fabric in

> Warp way : 44 lbs
>
> Weft way : 25 lbs

- Tensile strength of enzyme treated fabric is significantly higher than the normal treated fabric

7.5.1.6   Comparison of pH

This table 7.6 shows the comparison of pH value between the normal and enzymatic process.

Table 7.6

Type	Normal process	Enzymatic process
Desizing:		
pH	7.3	7.25
Scouring:		
pH	11.5	10.8

Most enzymatic activities are extremely sensitive to the pH and its variations. A variation of pH during the course of reaction may bring about an alteration in the protein structure with a denaturing effect on enzymatic or the ionization of the active site. The optimum pH also varies from one enzyme to another. In order to achieve a high yield from the enzymatic catalyst, it is important to adjust the pH to an optimum value. The table shows the pH value of the normal and enzymatic process.

7.5.1.7   Dye strength comparison

The dye strength comparison is measured in terms of the affinity to fibre and dye.

For cotton woven fabric

$\Delta l = 0.74$

$\Delta a = 0.52$

$\Delta b = 0.36$

$\Delta E = 0.97$

- The enzyme treated sample is 5% darker than the normally processed sample.

For cotton knitted fabric

$\Delta l = 0.86$

$\Delta a = 0.32$

$\Delta b = 0.36$

$\Delta E = 0.98$

- The enzyme treated sample is 10% darker than the normally processed sample.

7.5.2   Comparison of the properties of normal process and enzymatic process

Cotton Woven fabric

This table 7.7 (a) shows the comparison of properties of woven fabric between the normal and enzymatic process.

Table 7.7 (b): Cotton

Properties	Normal process	Enzymatic process
Absorbency(secs)	7.2	4.8
Whiteness index	63	73
Tensile strength(lbs)	34	44
Wash fastness	3.4	4
Rubbing fastness	4	4.5

Cotton Knitted fabric

This table 7.7 (b) shows the comparison of properties of knitted fabric between the normal and enzymatic process.

Table 7.7 (b)

Properties	Normal process	Enzymatic process
Absorbency(secs)	4.2	2.8
Whiteness index	63	73
Bursting strength(lbs/sq.inch)	50	87.73
Wash fastness	3.1	4
Rubbing fastness	4	4.5

7.5.3   Finishied fabric test results

7.5.3.1   *Tensile strength test*

Cotton Woven fabric

Table 7.8

Sample No	Direction	Normal softened fabric(lbs)	Enzymatic softened fabric(lbs)
1		28	40
2		30	40
3		33	36
4		29	38
5	Warp	31	36
6		25	38
7		26	36
8		26	34
9		28	36
10	Weft	26	37

Tensile strength of normally softened fabric in
Warp way : 30lbs
Weft way : 26lbs
Tensile strength of enzymatic softened fabric in
Warp way : 38lbs
Weft way : 36lbs

7.5.4   Comparison of effluent characteristics

Process	Normal process			Enzymatic process		
	pH	TDS	BOD	pH	TDS	BOD
Desizing	7.3	8900	4500	7.25	9000	4500
Scouring	11.0	5500	2140	10.8	1700	500
Bleaching	10.8	2390	500	10.8	1700	180

7.6   Conclusion

It is concluded that through this project, the cotton fabrics were processed using enzymes. The bio-processing is carried out by using amylase, pectinase, and catalase enzymes. The cotton fabric is treated with pre-treatment like desizing, scouring and bleaching using enzymes namely amylase, pectinase, and catalase. The required conditions were maintained as per the standard.

- The whiteness Index of enzymes treated fabric (73) is more than normal processed fabric (63).
- The colour strength of fabric is increased in the enzyme treated fabric. In woven cloth, the depth of colour is increased up to 5% and 10% is increased in knitted cloth.
- The washing and staining fastness of bio-processed fabric is found and it is noticed that slight increase in fastness properties.
- The pH values of the effluents of both bio-processing and chemical processing were found and it was noticed hydrogen ion concentration of bio-processing effluent is neutral for desizing and scouring.
- The tensile strength of the enzyme treated fabric is more compared to normal processed fabric.
- The hairiness of bio-washed fabrics are very less compared to normal processed fabrics.
- The enzyme treated fabric bursting strength is double, when compare to the normal processed fabric.
- Degradation of cellulosic materials is less, especially for full whites where high peroxide concentration is used.
- The effluent produced in the bio-processing is very less. So it will increase the life time period of effluent treatment plant and more water is processed in less time.
- In the case of cotton woven fabric the enzymatic processing cost/Kg is 11% lower than the normal processing.
- In the case of cotton knitted fabric the enzymatic processing cost/Kg is 7.5% lower than the normal processing.

7.7   Reference

1. M.Muthu Manickam, Application of Bio-technology in Textiles, p. 41-48, Colourage, Oct 2005.

2. Grosicki Z.J. *"Watson's Advanced Textile Design."* Newnes – Butterworths Universal Publishing Corporation., London., June 1999, PP 1-30

3. Shenai V.A. *"Technology of Textile Processing"* Vol VI, Sevak Publications, Bombay, 1998, PP 174 – 200

4. Goswami B.C and Martindale J.G." *Textile Yarns"* John Wiley & Sons. New York. 1977, PP 262 – 272.

5. Joseph Marjory L.., "Textiles - Weave determination for Section XI of the Harmonized System.," *Introductory Textile Science*, Vol. NHM 004, 1997, PP. 46-50.

6. K.N. Ninge gowda, Enzymes in Textile Industrial Applications, p. 15-19, Journal of the Textile Association, June-2004.

7. Geeta N. Sheth, Single bath Bio-scouring and Bleaching of Cellulosic yarn, Knitted and Woven fabrics, p.49-52, Colourge, November 2005.

8. S. R. Shukla, Production and Application of cellulose enzyme, p.41-46, Colourage, july 2005.

9. Dr. J. Jeyakodi Moses, "Colours produced from Natural dyes suitable for living environment", Indian Textile Review 2003, Vol.29 (2 nd Half yearly issue), 2003.

10. Dr. A.Venkatachalam & Dr. J. Jeyakodi Moses, "Convention proceedings on Natural Dyes", IIT, Delhi, p 121-130, Dec.17-18, 2001.

11. S. Gupta & M.Chakraborty, "Some Basic Concepts of Eco Textile and status of Banned Azo Dyes", Colourage and Apparel, March 2001.

12. Dr. A. A. Ansari & Dr. B. D. Thakur, "Extraction, Characterisation & Application of a Natural Dye: The Eco-friendly Textile Colorant", Colourage, p.5-15, July 2000.

8

Optimization on eco-friendly crease resistant finishing of silk fabrics

Dr. M. Parthiban* & Dr. M. Rameshkumar** & Dr. K. Saravanan***

*Assistant Professor(SG), Department of Fashion Technology,
PSG College of Technology, Coimbatore-641004*

**Associate Professor, Department of Fashion Technology,
Sona College of Technology, Salem - 636005, Email :rameshkumartex@gmail.com*

***Professor & Head, Department of Fashion Technology,
Bannari Amman Istitute of Technology, Sathyamangalam-638 401*

Abstract : The main objective of the research work is to

(i) Compare the wrinkle-resistant behavior of silk treated with conventional and formaldehyde free chemicals.

(ii) Understand the chemistry behind the action of new crease resistant finishing agents

(iii) Suggest an optimized process for effective wrinkle free finishing of silk fabrics by considering ecology and economy.

Keywords : Crease recovery angle, Eco-friendly finishes, Fastness, Silk & Tensile strength

8.1   Introduction

Silk fabrics have low wet and dry resiliency. Hence the fabrics wrinkle easily during home laundering or when wet[1, 2]. To improve these performance properties, silk fabrics are given chemical treatment known as durable press finishing. Durable press chemical finishes applied to silk fabrics in the presence of appropriate catalyst impart wrinkle resistance and smooth drying properties[3, 4]. The release of formaldehyde vapors is another problem with those agents. The most likely used cross-linking agents increase resistant finishes have been N- Methylol agents or N- Methyl amides because of their efficiency and low price. Formaldehyde free cross-linking agents for producing crease resistant properties are of interest to replace N-methylol compounds for crease resistant finishes [5,6].

Polycarboxylic acids which are non-formaldehyde reactants are a possible replacement for conventional finishing reactants. The main advantage of polycaboxylic acids is that they are formaldehyde free, do not have a bad odour, and produce very soft fabric hand [7,8]. Based on the above premise, an attempt has been made to try to assess the effect of polycarboxylic acids on silk with respect to its crease resistance behavior.

8.2   Materials and methods

8.2.1   Materials and their specification

8.2.1.1   Fabrics

Two different kinds of silk fabrics namely Mulberry and Tassar Silk were used for the experiment.

8.2.1.2   Chemicals used

   (a) Cross-linking agents: Glyoxal, Citric acid, and DMDHEU
   (b) Catalysts used: Aluminum Sulphate, Magnesium Chloride & Sodium Hypophosphate
   (c) Softener

8.2.2   Methods

8.2.2.1   Degumming of silk

Silk fabrics were first degummed using soap (8 gpl) for two hours at a temperature of 90°C after which they were washed and dried. They were then treated with the cross-linking agents as explained below:

8.2.2.2   Application of glyoxal

Both the fabrics were treated with three different concentrations of glyoxal viz, 5%, 10% & 15% (owf). Recipe for padding bath is as follows:

Glyoxal 5%, 10%, 15% (owf)

Aluminium Sulphate 3% (owf)

Softener 2% (owf)

Liquor Ratio 1:20

Temperature: Room Temperature

The fabrics were padded for about half an hour and then passed through the padding mangle for 10% expression. The fabrics were then tried and cured at 120°C. They are then washed and dried.

8.2.2.3   Application of citric acid

Both the fabrics were treated with 4 different concentrations of citric acid viz, 6%, 8%, 10%

& 15% (owf).

Recipe for padding bath is as follows:

Citric Acid: 6%, 8%, 10% & 15% (owf)

Sodium hypophospate 6% (owf)

Softener 2% (owf)

Liquor Ratio 1:20

Temperature: Room Temperature

The fabrics were padded for about half an hour and then passed through the padding mangle for 10% expression. The fabrics were then tried and cured at 120°C. They are then washed and dried.

8.2.2.4   *Application of Glyoxal*

Both the fabrics were treated with two different concentrations of DMDHEU viz, 6% & 10% (owf)

DMDHEU 6% &10% (owf)

Magnesium Chloride 6% (owf)

Softener 2% (owf)

Liquor Ratio 1:20

Temperature: Room Temperature

The fabrics were padded for about half an hour and then passed through the padding mangle for 10% expression. The fabrics were then tried and cured at 120°C.. They are then washed and dried.

8.2.2.5   *Dyeing*

The fabrics were dyed by using acid dye and then washed and dried at room temperature.

Recipe for dyeing bath as follows:

Dye 4% (owf)

Acetic acid 5% (owf)

Glauber's salt 10% (owf)

Liquor Ratio 1:40

Temperature: Room Temperature

Time: 1 hour

8.3   Testing methods

8.3.1   Sample preparation

Samples from both the fabrics were taken and the following tests were conducted:

Ambient testing conditions: RH 65 + (or) 2%; Temperature 25 + (or) 2°C

8.3.2   Crease recovery

Crease Recovery is quantitatively measured in terms of crease recovery angle using Eureka crease recovery tester. The sample size taken for testing was 2 × 1 cm

8.3.3   Tensile strength

The tensile strength was determined using Eureka tensile strength tester. The gauge length selected was 20cm x 5cm and ravel strip method was adopted.

8.3.4   Abrasion resistance

The abrasion resistance was determined using Martindale abrasion tester. In the instrument, a multidirectional movement is given to the fabrics, which in turn is mounted on the top plate, and abraded against emery paper.

8.3.5   Washing fastness

The washing fastness was determined using Wash Fastness Tester. In the instrument, the fabrics are treated with 5% soap solution at 40°C.

8.3.6   Rubbing fastness

An adequate number of pieces of 5 × 5 cm undyed bleached cotton were rubbed on dyed material of size 14 × 5 cm using crock meter. They were rubbed about 10 seconds with a downward force of 900g on the finger.

8.4   Results and discussion

The results of various treatment and test have been tabulated and they are discussed suitably and the suitable conclusion has been drawn.

8.4.1   Crease recovery angle

The crease recovery angles of control and treated Mulberry and Tassar fabrics are presented in Table 4.1 and depicted graphically in Fig 4.1(a), 4.1 (b) and 4.1 (c) respectively. There is a significant improvement in the crease recovery angle with all three treatments. The crease recovery angle increases from 130 degree to 141 degree with glyoxal, from 130 degree to 140 degree with citric acid treatments. The DMDHEU although shows an increasing trend, the increase is much lower compared to the other two methods. The maximum increase is observed with 5% of glyoxal, 6% of citric acid, and 10% of DMDHEU. It is observed that as the concentration of chemical increases the crease resistant angles decreases significantly. When the concentration of citric acid was lower the crease recovery angle found to be increased with increasing concentration of citric acid, this was due to the increase in the cross-linkages between silk polymer molecules. When the concentration was too high, the finish reacted too severely with the fibre and formed a thick layer on the surface of the fabric. This can reduce the resiliency and increases the specific density of the fabric. For the same reason, too much finish reduced the whiteness of the fabric. Tassar exhibits a remarkable improvement in the crease recovery properties with all the treatment as compared to mulberry. The increase in crease resistance is much higher at 81 degrees (glyoxal), 86 (citric acid) and 91 degrees DMDHEU from 71 degrees (untreated samples). The decrease in crease recovery angle with an increase in concentration is observed here. This may be due to the difference in the basic structure and constituents of tassar and mulberry fabric, and these fabrics have greater interlacing points per unit area.

8.4.1.1   Effect of catalyst

The action of the above eco-friendly chemicals enhance by the use of catalyst. In the present study two catalyst i.e., Aluminum Sulphate and Magnesium Chloride were used. The addition of aluminium sulphate catalyst and softeners to finishing bath can help in full swelling of silk fabric and assist the penetration of finishing agent into the fibres. Therefore, that cross-linking positions are increased resulting in a considerable increase in the crease recovery angle. When magnesium chloride is used as a catalyst for applying 5%, concentration of glyoxal the crease recovery angle is reduced from 135 to 115 degree. This is due to the presence of magnesium chloride catalyst and less fabric weight gain, besides it results in fabric yellowing. There may be some effect on the formation of cross-linking also.

8.4.1.2   Effect of curing temperature and time

At higher curing temperature the increase in stiffness, yellowness, and decrease in whiteness index results with irrespective of the finishing conditions used. This shows that higher temperature and long curing time are not favorable. In order to improve the resiliency and wash durability with a complimentary limited scarifies to other properties of silk, curing temperature at 120 to 130 degree with curing time of 120 seconds found to be the most suitable one.

8.4.2   Tensile strength

The tensile strength values of control and treated mulberry and tassar fabrics are presented in Table 4.2 and are depicted graphically in Fig 4.2(a), 4.2 (b) & 4.3 (c). There is a significant decrease in the tensile strength, which was found in all the three treatments. In the case of Mulberry, tensile strength decreases from 45 kg to 41 kg with 5% glyoxal, from 45 kg to 40 kg in 6% citric acid treatments. No doubt, this substantial strength retention was found after the treatment. There is only a 12% loss in tensile strength results in fabric, which are treated with glyoxal and citric acid at 5% and 6% concentrations respectively, with increase in concentration and decrease in tensile strength results. This may be due to acid catalyzed depolymerisation during the curing process. The magnitude of fabric strength loss may be affected by the pH of the acid solution applied to the fabric and the cross-linking system becomes inefficient when the pH is increased to 3.65. However, a 12% loss in tensile strength is not an acceptable value to take into consideration. In the case of Tassar fabric, significance difference in tensile strength loss was observed with all the three treatments with different concentrations. From the results, it is observed that the tensile strength drops from 35 Kg to 10 Kg with 5% glyoxal and 35 Kg to 4 Kg with 6% citric acid. The considerable decrease in tensile strength was also found in DMDHEU. It is observed that when the concentration of chemical is increased, the tensile strength suffers further loss. This is due to embrittlement and molecular degradation of tassar fabric.

8.4.3   Abrasion resistance

The weight loss percentage of control and treated mulberry and tassar fabrics are presented in Table 4.3 and are depicted graphically in Fig 4.3(a), 4.3 (b) & 4.3 (c). In the case of mulberry, there is a significant increase in weight loss percentage with all the three treatments, the weight loss percentage increases from 3% to 5% with 5% glyoxal and 6% citric acid and 10% DMDHEU. It is observed that as the concentration of chemicals are increased the weight loss percent increases i.e., abrasion resistance decreases. The abrasion resistance is affected largely at 15% concentration of glyoxal and citric acid. This may be due to abrasion resistance being associated with the tensile strength, which in turn is associated with the extent of cross-linking. However, additives may have an important role to play on this property. In case of Tassar, there is no significant decrease in the weight loss percentage of fabric treated by 5% glyoxal, 6% citric acid and 10% DMDHEU. However, increase in concentration will result in increase in weight loss percentages i.e., decrease in abrasion resistance. Abrasion resistance is affected greatly to the extent at 15% concentration of glyoxal, 15% concentration of citric acid and 10% concentration of DMDHEU. This may be due to abrasion resistance beign associated with tensile strength which in turn is associated with cross-linking. However, additives may have an important role to play on this property.

8.4.4   Dye uptake behaviour

The fabrics were dyed with an acid milling dye. A significant difference was found visually in dye uptake of treated and controlled Mulberry and Tassar fabrics. Much work was not carried on this due to lack of time and available facility.

8.4.4.1   Washing fastness

The fabrics treated with 5% glyoxal (aluminium sulphate), 6% citric acid, 0% DMDHEU and controlled were tested for wash fastness, the grey scale rating of fabrics was found to have 4 for fabrics treated with 5% glyoxal (aluminium sulphate), 6% citric acid and 4/5 for untreated fabric 10% DMDHEU treated fabric.

8.4.4.2   Rubbing fastness

The fabrics treated with 5% glyoxal (aluminium sulphate), 6% citric acid, 10% DMDHEU and controlled were tested for Rubbing Fastness, the grey scale rating of fabrics were found to have 4 for fabrics treated with 5% glyoxal (aluminium sulphate), 6% citric acid, 10% DMDHEU treated fabric and 4/5 for untreated fabric.

8.5   Conclusion

The optimum concentration for treatment with glyoxal and citric acid were found to be 5% and 6% respectively. The aluminum sulphate was found to be the most suitable catalyst for glyoxal. The optimum curing temperature and curing time were found to be 120°C and 120sec. Eco-friendly crease resistant finishes were found to a have less deleterious effect on mechanical properties i.e., tensile strength, abrasion resistance, etc., as compared with DMDHEU. The percentage of increase in crease recovery angle was found to be more in Tassar fabric than with Mulberry fabric. The work can be extended by using different catalysts and different crease resistant agents. Formaldehyde release can be extensively studied to know about the nature of various finishes. It can also be extended on other fabrics like cotton, viscose, etc.

8.6   References

1. Andrew, Wrinkle Resistant cotton and formaldehyde release, Colorage, 41, 87-93 (1995).

2. Dr S A Shah and K S Taraporewala, Manmade Textile in India (1995).

3. Reinhardt R M Bhattacharya N Doshi B A Sahasrabuddha A S and Ministry P R , Citric Acid treatment of dyed cotton fabrics, American Dyestuff Reporter, 84, 17- 20 (1995)

4. Reinhardt R M Bhattacharya N Doshi B A Sahasrabuddha A S and Ministry P R, A Comparison of BTCA and DMDHEU Cross- linking treatments on dyed cotton fabrics American Dyestuffs Reporter, 83, 80 -90 (1994).

5. Welch C M and Peters J G Mixed Polycarboxylic acid mixed catalyst in formaldehyde free durable press finishing, Textile Chemist and Colorist, 29, 22 -72 (1997).

6. Brodmann G L, Performance of non-formaldehyde release from durable press fabrics, Textile Chemist and Colorist, 14, 100 -106 (1982).

7. Welch C M and Andrews B A K. Ester crosslinks: A route to high performance non-formaldehyde finishing of cotton, Textile Chemist and Colorist, 21, 13 -17(1989).

8. Peterson H, The chemistry of crease resistant crosslinking agents, Review of progress in coloration and related topics, 17, 7-22 (1987).

9

Effect of repllent finishing of fabrics using vetiver extract

Dr. M. Parthiban* & Dr. M. R. Srikrishnan*

*Department of Fashion Technology, PSG College of Technology, Coimbatore 641 004, Tamil Nadu, India, Email: parthi_mtech@yahoo.com

Abstract : Insects play a vital role in the transmission and spread of any disease not only in rural but also in urban areas. Insect repellents have been used to reduce human-vector contact for long periods in different parts of the world protecting the human beings from the sting of mosquitoes and other insects thereby promising safety from the insect-carried diseases. Insect repellent textiles are one of the revolutionary ways to promote the textile field by providing the most demanded features of driving away mosquitoes and other insects. The main objective of this study is to assess the performance of the insect repellent products developed from Vetiveria zizanioides against mosquitoes and other insects. A questionnaire-based study has been carried out and the results revealed that commercially available insect repellents are harmful to human health and their use should be avoided. It is revealed that the developed insect repellent products using plasma treated and vetiver powder provided a solution to the major problem faced by the society.

Keywords : Insect repellents, Vetiveria zizanioides, Mosquito Repellent Textiles, Mosquito repellent finishes, Results & Discussion.

9.1 Introduction

Today's era is to bring in modernization to the textile industry. Every sector of textile and every field related to textiles is developing with the advancements taking place. Smart textiles or functional textiles are one such field. This field is the fastest growing one in the textile industry. Protective textiles are among one such smart application of smart technology in textiles. Protective textiles refer to those textile products, which have a functionality of giving protection from something in some or the other sense. These can be mosquito repelling or may be insect repelling and also maybe anti-bacterial and antifungal. These may also be heat and cold resistant or with any other property. Although this sector contributes 1% to the total sales of the smart textiles, still has a good scope for growth.

Vetiver grass (Vetiveria zizanioides) also known as chrysopogom zizanioides, is a graminaceous plant native to tropical and subtropical India. In western and northern India, it is popularly known as khus. Vetiver is the most versatile, multifarious grass with immense potential. Vetiver is fibrous and its aromatic roots have been harvested for centuries and turned into perfumes, insect-repelling textiles, closet sachets, and even food flavorings. Vetiver root paste or its extract is used as a diaphoretic, stimulant, and refrigerant, flatulence and obstinate vomiting. It is a plant known for its ability to produce essential oil from the roots, which are especially used in the perfume industry. There is increasing interest in the health and wellness benefits of herbs and botanicals, this is with good reason as they might offer a natural safeguard against the development of certain conditions and be a putative treatment for some diseases. However, from ancient times, Vetiveria zizanioides have also been used as raw materials for cosmetics, pharmaceuticals, botanical pesticides, disinfectants, insect repellents, herbal teas, etc. Growing demands on the functionality of textiles as well as the environmental friendliness of the finishing processes increase the interest in physically induced techniques for surface modification and coating of textiles.

9.2   Repellant finishes

Basically, mosquito repelling textiles are the ones which have a character of repelling mosquitoes. This feature was developed as a need in sense of protection from the mosquitoes in the areas which are habitats of the mosquitoes and are prone to disease like malaria. To impart this feature, the textile material is given an anti-mosquito finish with an agent. This agent is capable of being used on textiles without spoiling their characteristics and has good washing fastness.

9.3   Significance of repellant finishes

Because of global warming, the distribution of mosquitoes has expanded from tropical regions to northern latitudes, and that leads to a spread in sources of viral infection from mosquitoes. Especially, the West Nile fever virus, which has infected many people around the world recently, has become a big issue. West Nile fever occurs routinely in Africa as its virus was first recognized in humans at the West Nile District of Uganda in 1937, and then infections were confirmed in Israel, France, and South Africa and it is now showing signs of spreading further all over the world. The first outbreak occurred in New York City in 1999, which spread rapidly to over 4000 people all over the U.S. and killed over 240 people. It is likely that the virus arrived via wild birds imported as pets and via plane or boat in an infected mosquito. Persons diagnosed with the disease have recently even been confirmed in the island country of Japan with the wave of globalization, and the Ministry of Health,

Labour and Welfare requires cases to be reported to prefectural governors under regulations for disease control and prevention.

9.4   Mechanism of repellency

The action of repellent agent for blood-sucking insects including mosquitoes can be broadly divided into two types which are actions to repel insects by acting on the olfactory and tactile senses. The action of repellent on the sense of smell is called *transpiration repelling*, and this has the effect of keeping insects away without them touching a surface processed with the repellent agent. How it works is that repellent molecules block insect's humidity sensory holes, which makes humans inaccessible to insects by inhibiting the function of sensing moisture, while insects usually use warm and humid convection rising from the human body as a guide for contacting humans sensing an increase in atmospheric carbon dioxide concentrations. The action of repellent stimulating sense of touch is called *direct-contact repelling*, and this drives insects off the processed surface before blood sucking even after touching the surface. It is believed that repellent substances work on insects peripheral nervous systems when contact is made, causing a collaterally-expressed confusional state and inhibition under sub-lethal doses before knockdown and lethal action.

9.5   Objectives

- To investigate the effect of oxygen plasma as a pre-treatment to the herbal finishing;
- To develop anti-bacterial and mosquito repellent textile products with selected herbs;
- To investigate the effect of herbs on the mosquito repellency of textiles;
- To assess the mosquito repellent activity of finished fabrics;
- To convert the finished fabrics into consumer products and evaluate them for consumer acceptance.

9.6   Materials & methods

9.6.1   Materials

The present study was conducted to assess the performance of the developed insect repellent products against mosquitoes and other insects. 100% organic cotton and vetiver root powder has been selected for the study.

9.6.2   Methods

This includes a coating of vetiver solution extracted from the methanol preparation after 24 hours on the plasma treated fabric using atmospheric oxygen followed by pad dry cure method. 1:10 ratio of vetiver powder in grams to methanol solution in ml has been dissolved and kept in a beaker for 24 hours respectively and the extracted solution has to be filtered for any further use. The 20 × 20cm fabric has been treated with atmospheric oxygen plasma under 0.01mg pressure for 30 seconds. The process has been continued for N number of samples and it has been followed by pad -dry cure within 4 hours from plasma treatment. The plasmatreated fabric and the untreated fabric has kept in the extracted solution of vetiver for 3-4 hours respectively and the cured by padding mangle machine to evenly spread out the solution on the fabric.

9.6.3   FTIR

Fourier Transform Infrared Spectroscopy, also known as **FTIR Analysis** or **FTIR** Spectroscopy, is an analytical technique used to identify organic, polymeric, and in some cases, inorganic materials. The **FTIR analysis** method uses infrared light to scan **test** samples and observe chemical properties.

9.6.4   GC-MS

Gas chromatography-mass spectrometry (GC-MS) is an analytical method that combines the features of gas-chromatography and mass spectrometry to identify different substances within a test sample. Applications of GC-MS include drug detection, fire investigation, environmental analysis, explosives investigation and identification of unknown samples.

Gas column chromatography parameters:
- Equipment : thermo gc - trace ultra ver: 5.0, thermo ms dsq ii
- Column   : db 35 - ms capillary standard non - polar column
- Dimension : 30 mts, id : 0.25 mm, film : 0.25 μm
- Carrier gas : he, flow : 1.0 ml/min
- Temp prog : oven temp 70 c raised to 260 c at 6 c/min

9.7   Testing standards

9.7.1   Repellency test (Excito chamber method)

The mosquito repellency efficiency of the developed nonwovens was tested using the modified Excito chamber method. There have been numerous attempts to accurately measure the behavioral responses of mosquitoes to the fabrics.

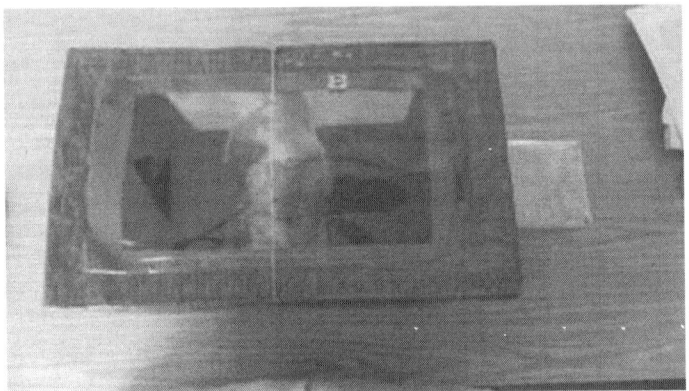

Plate 1: *Excito Chamber*

9.7.2   Cone bioassay test

Cone test of an insecticide-treated textile. Mosquitoes (usually malaria mosquitoes, *Anopheles gambiae*) are introduced into a standardized cone for a defined time span, then removed and transferred to small cages to determine the knock-down and knock-dead rates. The test samples with the size of 20cm were placed in glass containers with ten larvae of the target species. After 3 minutes of exposure, two observations were made at 1 hour and 24 hours respectively. Percentage mortalities were made at the end of 24 hours.

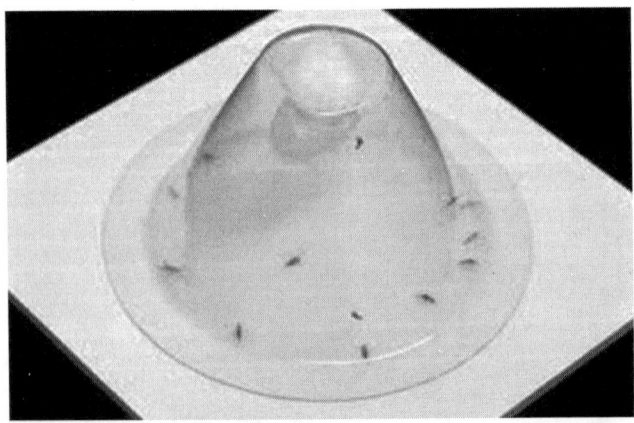

Plate 2: *cone bioassay test*

9.8   Results and discussion

9.8.1   Characterisation studies of vetiver extract on fabrics

9.8.1.1   FTIR

The plasma treated sample coated with vetiver solution extracted by methanol followed by pad dry cure method has been tested under Fourier transform infrared spectroscopy and the following compound name has been detected.

- Cellophane
- Chipboard (K540 4.2% N , P40 10.7% N , K540 2.9% N)
- Cellulose+lignin
- Wood+melamine-formaldehyde resin
- Chipboard w/ 3.6% methylene bis (phenyl isocyanate)
- Chipboard w/ urea-formaldehyde condensate

Cellulose and cellophane presence indicates that the treated fabric is pure cotton. All the other compounds indicate the presence of vetiver chemical components that actually has the characteristics of killing or repelling of mosquitoes.

9.8.1.2   GC-MS

The vetiver residue was composed of

1. Hydrocarbons
2. Alcohols
3. Ethers
4. Ketones
5. Esters
6. Sesquiterpenes processing cedrane, bisabolane, eudesmane, eremophilane & Zizaane skeletons.

The major components present in the treated fabric are

- Eudesma 4, 6-dine+beta vetispirene = (3.9-6.1%)
- Beta vetivenene = (0.9-9.4%)
- 13-nor-trans-eudesma-4(15), 7-dien11-one+amorph 4 –en-10-ol = (5-6.4%)
- Trans-eudesma-4 (5), 7-dien-12-ol (vetiselinenol) + (E)-opposita-4(15), 7(11)-dien-12-ol = (3.75.9%)

- Eremophila-1(10), 11-dien-2a-ol (nootkatol) +ziza-6(13)-en-12-ol (khusimol) = (16.1-19.2%)
- Eremophila-1(10), 7(11)-dien-2a-ol (isonootkatol) + (E)-Eremophila-1(10), 7(11)-12-ol (isovalencenol) = (5.6-6.9%)

So overall mosquito repellency percentage of plasma treated fabric:

- Alpha vetivone and beta vetivone were found to be repellent to insects and their overall percentage level from the mass spectrum values is 124.9%.
- The percentage level of mosquito repellency of vetivone derivatives is 58.97%.
- The percentagelevel of mosquito repellency of other derivatives is 16.1-19.2%
- The overall percentage level of mosquito repellency is 75.07-78.17%

9.8.2   Performance of repellant finishes on fabrics

Sample 1: Vetiver treated fabric, blue color

Table 9.1: Excito chamber repellency test

Sample Identification	No. of Mosquitoes released in Chamber	No. of Mosquitoes on Treated Fabric	No. of Mosquitoes Showing mobility	Percentage repellency
Sample 1 – Blue Color (Initial)	10	10	00	00
Sample 1 – Blue Color (½hr)	10	10	30	30

P.S.: Percentage Repellency Efficiency is equivalent to Percentage protection imparted by the fabric.

Sample 1, fabric treated with vetiver solution followed by pad dry cure method showed only 30% repellency under excito chamber test method.

9.8.3   Effect of repellent finishes on plasma treated samples

Sample 2: Vetiver treated fabric, blue colour. **Sample3:** Plasma come vetiver treated fabric, dirty white color.

Table 2: Cone bioassay repellency test

Sample Description	No. of Mosquitoes Released	Observations on-Number of Mosquitoes Knocked down		Percentage Mortality
		after 1 hour	after 24 hours	
SAMPLE 2: Blue Colour, Untreated	10	00	04	40
SAMPLE 3: Dirty White Colour, Plasma Treated	10	01	07	70

Sample 2 showed only 40% repellency whereas **sample 3** showed 70% repellency, showing that the plasma-treated fabric has the best repellency than the normally treated fabric here. It shows the plasma treated fabric has good bonding affinity towards the coated vetiver solution by changing its surface modification respectively.

9.8.4   Effect of repellant finishes on durabililty of the fabrics

Sample Description	No. of Wash Cycles	No. of Mosquitoes Released	Observations on-Number of Mosquitoes Knocked down		Percentage Mortality
			after 1 hour	after 24 hours	
Sample 2 -Blue Color, Untreated	5	10	00	00	00
Sample 3-Dirty White Color, Plasma Treated	10	10	00	04	40
Sample 3 -Dirty White Color, Plasma Treated	15	10	00	00	00

The wash cycle test results shows that the **Sample 2**, which is normally treated with vetiver extraction followed by pad dry cure method showed no killing/repellency after first 5washes, whereas the sample3, which is plasma treated along with vetiver extraction followed by pad dry cure method withstands over 10 washes showing 60% repellency in the 5th wash and 40% repellency in the 10th wash respectively. It shows the plasma treated fabric has good durability over wash cycling process than the normally treated fabric.

9.8.5   Effect of plasma treated fabric vs repellency

In theoretical testing, the plasma-treated fabric showed 75-78% of repellency overall whereas, in practical mosquito repellency testing, the plasma-treated fabric showed 70% of repellency. Hence, it is proved that it has been balanced.

9.9   Conclusions

Many researchers report that there were lots of problems and harmful side effects due to the commercially available insect repellents. People were ready to buy herbal repellents which are completely safe and they were not ready to use easily available chemical based repellents. From this study, it was concluded that Vetiveria zizanioides were the natural and herbal finishing agent, which were applied to the plasma treated fabric using pad-dry-cure-technique, have very good mosquito repellent finish. The natural finishing agents were eco-friendly bio-degradable, non-toxic, and non-irritant to the skin. It shows good mosquito repellent property when applied on the plasma treated cotton fabric than normally treated vetiver fabrics. They were evaluated by excito chamber test, cone test. The present study was carried out based on the survey report using 100% organic cotton and vetiver solution along with plasma treatment for better durability and it was found that sample3 (plasma treated-Vetiver solution) was good in all aspects. These forms of natural insect repellent products were very safe and eco-friendly and protect us from mosquitoes and other insects.

9.10   Future Scope

The use of commercial insect repellents such as mats, lotions, coils, refills, hits, and liquidators are limited due to various health problems. Insect repellency using natural herbs is the much-needed feature of driving away the mosquitoes and other insects thereby promising safety from the insect-borne diseases. Efforts should be made in the future to promote the use of other herbal insect repellent plants to maintain the ethnobotanical knowledge of the inhabitants.

9.11   References

1. Tolle M A, Current problems in pediatric and adolescent health care, Apr 09, 39.

2. http://www.cmete.com/vaccinations-en/diseases transmitted-by-insects

3. http://en.wikipedia.org/wiki/Insect_repellent

4. Phasomkusolsil S and Soonwera M, Comparative mosquito repellency of essential oils against Aedes -aegypti, Anopheles dirus and Culex quinquefasciatus, Asian Pacific Journal of Tropical Biomedicine, 2011,S113-S118.

5. Mishra Snigdha, Sharma Satish Kumar, Mohapatra Sharmistha, and Chauhan Deepa, An Overview on Vetiveria Zizanioides, Research Journal ofPharmaceutical, Biological and Chemical Sciences, Volume 4 Issue on 3-July-September 2013, 777-783.

6. htt://www.fibre2fashion.com/industryarticle/46/4560/home-textile-products-with-vetiveria-zizanioides1.asp

7. Wright CI, Van-Buren L, Kroner CI, Koning MM, Herbal medicines as diuretics: a review of the scientific evidence, Journal of Ethnopharmocology, Oct 2007,114(1).

8. Aarthi N and Murugan K, Effect of Vetiveria zizanioides L.root extracts on the malarial vector Anopheles stephensi Liston, Asian Pacific Journal of Tropical Disease, 2011,154-158.

10

Application and analysis of anti mosquito finishing textiles

P. Vinayagamurthi[1] & Dr. S. Kavitha[2]

[1]*PhD Scholar - Textile Technology, Research & Development Centre, Bharathiar University, Coimbatore – 641 046. Tamil Nadu, India, Email: vmcostume@gmail.com*

[2]*Associate Professor, Department of Home Science, Mother Teresa Women's University Research and Extension Centre, Coimbatore - 641 002 India, Email: s.kavithagiriraaj@gmail.com*

Abstract : In today's era of modernization of the textile industry, we are going through advancements of technology in every field of this industry. The world where this would lead us would be astonishingly hi-tech and materialistic. To ensure our security and safety from future hazards, we need to equally develop the technology for our protection. With regard to textiles, the protective textile field of the smart textiles has to fulfill this requirement. A Mosquito repellent textile is one such textile product. It protects human beings from the bite of mosquitoes and thereby promising safety from diseases like Malaria and Nile fever. To impart this character a finish of the mosquito-repelling agent is given to the textile material. Thorough research and development have facilitated the applicability of certain chemicals on the textile products, which sustain this character for a reasonable period. Insect repellent textiles are also a part of protective textiles, which help in protection from the species that are prone to cause damage in some, or the other manner. These textile products find their application over a wide range. A lot has been achieved and much more is yet to be covered, as there is no end to it. What can be done to the best is that this technology can be imparted to the fibres from within. In addition, keeping in mind the ecological and economical aspects, efforts should be made to make this all available to the commons. Mosquito repellent textiles are one of the most growing ways to advance the textile field by providing the needed characteristics of protection against mosquitoes, especially in tropical areas. These types of textiles ensure the protection of human beings from the mosquitoes and the mosquito-borne disease includes malaria, filariasis, and dengue fever. This study focused on the penetration of mosquito repellent finish in textile applications as well as nature-based alternatives to commercial chemical mosquito repellents in the market. Suitable technologies and materials to achieve mosquito repellency are discussed and pointed out their applications and further scope of research and development.

Keywords : Mosquito repellent, Anti-mosquito, Insect repellent, mosquito repelling agent.

10.1   Introduction

Mosquitoes transmit diseases worldwide to more than 700 million persons annually and account for 1 in 5 childhood deaths in Africa. Mosquitoes will kill 1 in 17 persons currently alive on the planet through diseases such as malaria, dengue, and mosquito-borne encephalitis. Mosquitoes traditionally have been a concern of the developing world, yet the arrival of West Nile virus in North America has renewed public attention and concern toward these insects. Indeed, the cost of medical care for mosquito-borne illness is rapidly rising in the United States. Many organizations for adolescents participate in summer-camp experiences, immersing themselves in the North American outdoors. Chief among these organizations is the Boy Scouts of America. In 2001, over 2 million youth members of the Boy Scouts of America participated in summer camp, placing themselves at increased risk for zoonotic diseases. With the rising concern over mosquito-borne arbovirus encephalitis, such as La Crosse or West Nile virus, avoidance of mosquitoes is the best protection against disease. There are mainly three varieties of mosquitoes, which spread most of the dreadful diseases among humans. They are *Anopheles, Culex, and Aedes.* The mosquitoes, which have the potential of spreading diseases among humans, are generally termed as **vectors**. Apart from these vectors, there are several species of mosquitoes, which are generally classified as *nuisance mosquitoes,* which may act as secondary vectors during epidemic periods. Many commercially available mosquito repellents contain N, N-diethyl-m-toluamide, now called N, N-diethyl-3-methylbenzamide (DEET). DEET is an effective topical mosquito repellent generally providing 1.5 to 5 hours of protection, depending on formulation and concentration. However, this DEET is a skin lotion and has to be applied to the skin, which may have some allergic response on some individuals. Furthermore, it is difficult to apply this finish on fabrics as their longevity is too short and hence will not be able to satisfy the requirement. In this work, an attempt has been made to provide mosquito repellent fabrics without any adverse effects to the users.

10.2   Mosquitoe species

Mosquitoes are found all over the world, except in Antarctica. These two-winged insects belong to the order Diptera. Members of the genera *Anopheles*, *Culex*, and *Aedes* are most commonly responsible for bites in humans.

Only female mosquitoes bite. Male mosquitoes feed primarily on flower nectar or plant sap, whereas female mosquitoes require a blood meal to produce eggs. They usually feed every 1 to 4 days; in a single feeding, a female mosquito typically consumes more than its own weight in blood[1]. Certain species of mosquitoes prefer to feed at twilight or nighttime; others bite mostly during the day. Some mosquito species are zoophilic (preferring to feed on animals) and others are anthropophilic (showing a preference for human blood). In some mosquito species, seasonal switching of hosts provide a mechanism for transmitting diseases from animal to human.

10.2.1   Stimuli that attract mosquitoes

The factors involved in attracting mosquitoes to a host are complex and are not fully understood[2-1]. Mosquitoes use visual, thermal, and olfactory stimuli to locate a host. Of these, olfactory cues are probably most important. For mosquitoes that feed during the daytime, movement of the host and the wearing of dark-colored clothing may initiate orientation toward a person[1, 4]. Visual stimuli seem to be important for in-flight orientation, particularly over long ranges, whereas olfactory stimuli become more important as a mosquito nears its host.

It has been estimated that 100 to 400 compounds are released from the body as by-products of metabolism and that more than100 volatile compounds can be detected in human breath[9]. Of these odors, only a fraction has been isolated and fully characterized. Carbon dioxide and lactic acid are the two best-studied mosquito attractants. Carbon dioxide, released mainly from breath but also from the skin, serves as a long-range airborne attractant and can be detected by mosquitoes at distances of up to 4 meters [1, 11-6]. Lactic acid, in combination with carbon dioxide, is also an attractant. Mosquitoes have chemoreceptors on their antennae that are stimulated by lactic acid. These same receptors may be inhibited by N, N-diethyl-1-methyl-benzamide (DEET)-based insect repellents.[4]

At close range, skin temperature and moisture serve as attractants[1,9,7]. Different species of mosquitoes may show strong biting preferences for different parts of the human body (such as the head or feet), which may be related to local skin temperature and eccrine sweat gland output[8, 10]. Anhidrotic people show markedly decreased attractiveness to mosquitoes[2]. Other volatile compounds, derived from sebum, eccrine and apocrine sweat, or the cutaneous microflora bacterial action on these secretions, may also act as chemoattractants[2, 11, 12]. Whole-host odors are more attractive than carbon dioxide and lactic acidalone[13]. Floral fragrances from perfumes, soaps, lotions, and hair-care products may also attract mosquitoes.[12]

The attractiveness of different persons to the same or different species of mosquitoes varies substantially[7, 14]. In general, adults are more likely to be bitten than children[7, 15], although adults may become less attractive to mosquitoes as they age [2]. Men are bitten more readily than women[1, 13]. Larger people tend to attract more mosquitoes, perhaps because of their greater relative heat or carbon dioxide output[17].

10.2.2   Extraction of pyrethrum from chrysanthemum

10.2.2.1   Water extraction

About 500 Gms of the flower is dried and then crushed to powder form. Then about 5 g of the powder is soaked in 100 ml warm water for about 2 hours. Finally, the supernatant is taken and the residue is thrown off. For extremely strong solution, concentrations the amount of water used may be reduced.

10.2.2.2   Kerosene extraction

About 500 Gms of dried flowers are soaked in 4 litres of kerosene for 4 hours. The kerosene will dissolve 71% of pyrethrin in 48 hours. After filtering through a fine mesh cloth, the extract is ready for application.

10.2.2.3   Paraffin wax

Dried flowers are soaked in paraffin for one day and occasionally stirred. The solution must be filtered before use.

10.2.2.4   Alcohol extraction

The procedure of extraction is the same as that of paraffin extraction. Alcohol extraction is possible but the resulting product has the disadvantage of evaporating very quickly.

Of all the extraction procedures, we have considered only water extraction procedure because of the following reasons. Kerosene extraction is not advisable, as it possesses harmful threats to the environment. Paraffin's and alcohols are highly volatile and hence not recommended.

10.2.2.5   Chemical Structure of Pyrethrum or Pyrethrin

Chemical Formula: $C_{19}H_{26}O_3$

Pyrethrum was initially derived from the crushed dried flowers of the daisy Chrysanthemum, whose insecticidal properties have been recognized since the 8th century.

10.3   Need for mosquito repellent fabrics

Until now mosquito repellents are used in the form of skin lotions and vaporizers. Usage of the repellents in these forms possessed several drawbacks. In particular, some of the repellent lotions used had serious allergic responses to certain individuals. The main drawback of using vaporizer repellents is that it cannot be used outdoors. In addition, one more serious problem with vaporizer repellents is that it is very harmful when inhaled; in particular, children are more susceptible.

In order to encounter the above-said drawbacks, it becomes imperative to find a solution that is much safer and reliable. It is a known fact that clothing acts as a second skin to the humans and from time immemorial; we have been using clothes to protect ourselves from the environment. However, imparting this finish on apparels does not sound meaningful hence, we have shifted our focus towards bed-spreads, curtains, and tents (used for camping purposes). So instead of using repellents as lotions or vaporizers, which possess a potential threat to the humans, the same property can be employed to the fabrics without posing any risks to human health.

This led us to take up this project so that a safe solution is presented for the problem caused by mosquitoes. Although several chemicals were available for this purpose, we wanted to provide a safe and natural way of imparting the repellent property. Pyrethrum was found to be one such compound, which has the ability to fulfill all the necessary requirements.

10.4   Materials and methods

10.4.1   Fabrics

For developing mosquito repellent fabrics initially, two fabrics were sourced. The first being cotton and the other sample being nylon. The logic behind choosing specifically these two fabrics is that cotton is widely used as bedspreads and nylon as mosquito nets and in camping and trekking gears.

In order to develop the repellent fabric, the fabric should be dipped in the chemical, which is extracted from the chrysanthemum flower. The method of extraction is as follows:

10.4.2   Mosquito spicies

The test mosquito species chosen for the study is *Aedesaegyptia* which is a primary vector for spreading the dreadful disease namely *Dengue*. The other species, which was chosen for testing, is *Arvigillus* which is a non-vector and is commonly referred to as nuisance mosquitoes. It is a strong variety and difficult to deal with.

10.4.3 Method of producing powdered extract

Initially, the flower is soaked in water for about 24 hrs and the soaked flower is dried in shade. Drying should be done only in shade, as exposure to sunlight will have detrimental effects on the amount of pyrethrum present in the flower. The dried flowers are then ground to fine powders using pestle & mortar. The powder is then sieved to remove larger particles and other macro materials so that a very fine powder results.

10.4.4 Extraction of pyrethrum

The finely powdered flowers of chrysanthemum contain the active ingredient namely the pyrethrum that has to be extracted. There are several solvents available for this purpose but some of them are volatile in nature and some are not suitable for its application in textiles as they are banned. The simple and the most effective method of extracting pyrethrum in terms of both cost and process is by using water. The powdered extract along with water is taken and the extraction is carried out at two different temperatures 60°C and 90°C.

About 5 g of the powdered extract along with 100 ml of water is taken. Two such baths are used. In one bath, the temperature is maintained at 60°C and in the other bath, 90°C is maintained. The process is carried out for two hours beyond which there is very little improvement in the concentration of the final extract solution. Out of the two baths, the bath maintained at 90°C showed poor quality of the final extract solution than that of the bath maintained at 60°C. This can be understood from the fact that raising the temperature beyond 70°C has a detrimental effect on the powdered extract.

Finally, the extraction is carried out at 60°C for about two hours. At the end of the process the final extract solution (is allowed to stand for at least two hours) is filtered off to remove the residues and the supernatant i.e., the final clear solution is used for the finishing purposes.

10.4.5 Finishing of fabrics

There are two methods by which the finish can be applied to the fabrics. They are

1. Padding method
2. Exhaust method

In the first method, the fabric is simply impregnated in the extract solution followed by drying and then finally cured. In the second method the fabric is treated with the solution for about an hour at 50°C-60°C and then it is squeezed and dried.

Both the methods were found to be effective to a large extent, in particular, the exhaustive method was highly successful for both nylon and cotton and less complicated than the padding technique. Nylon was finished only by the exhaustive method, as it was too difficult to maintain the parameters for padding technique. Since we used a mesh cloth for this purpose, it was not able to withstand the stresses and strains in the padding technique.

The following recipes were used for finishing.

10.4.6   Exhaust method

Recipe 1:

Pyrethrum extract	–	100%
M : L	–	1:30
Time	–	1 hr
Temperature	–	50°C

Recipe 2:

Pyrethrum extract	–	50%
M : L	–	1:30
Time	–	1 hr
Temperature	–	50°C

Recipe 3:

Pyrethrum extract	–	25%
M: L	–	1:30
Time	–	1 hr
Temperature	–	50°C

Recipe 4:

Pyrethrum extract	–	10%
M: L	–	1:30
Time	–	1 hr
Temperature	–	50°C

Recipe 5:

Pyrethrum extract	–	5%
M : L	–	1:30
Time	–	1 hr
Temperature	–	50°C

Recipe 6:

Pyrethrum extract	–	1%
M : L	–	1:30
Time	–	1 hr
Temperature	–	50°C

10.4.7   Padding method: [pad- dry- cure]

Recipe 1:

Pyrethrum	–	5%
Dextrin	–	500 parts
Water	–	500 parts
		1000 parts

Recipe 2:

Pyrethrum	–	10%
Dextrin	–	500 parts
Water	–	500 parts
		1000 parts

Recipe 3:

Pyrethrum	–	25%
Dextrin	–	500 parts
Water	–	500 parts
		1000 parts

10.4.8   Processflow chart for finishing

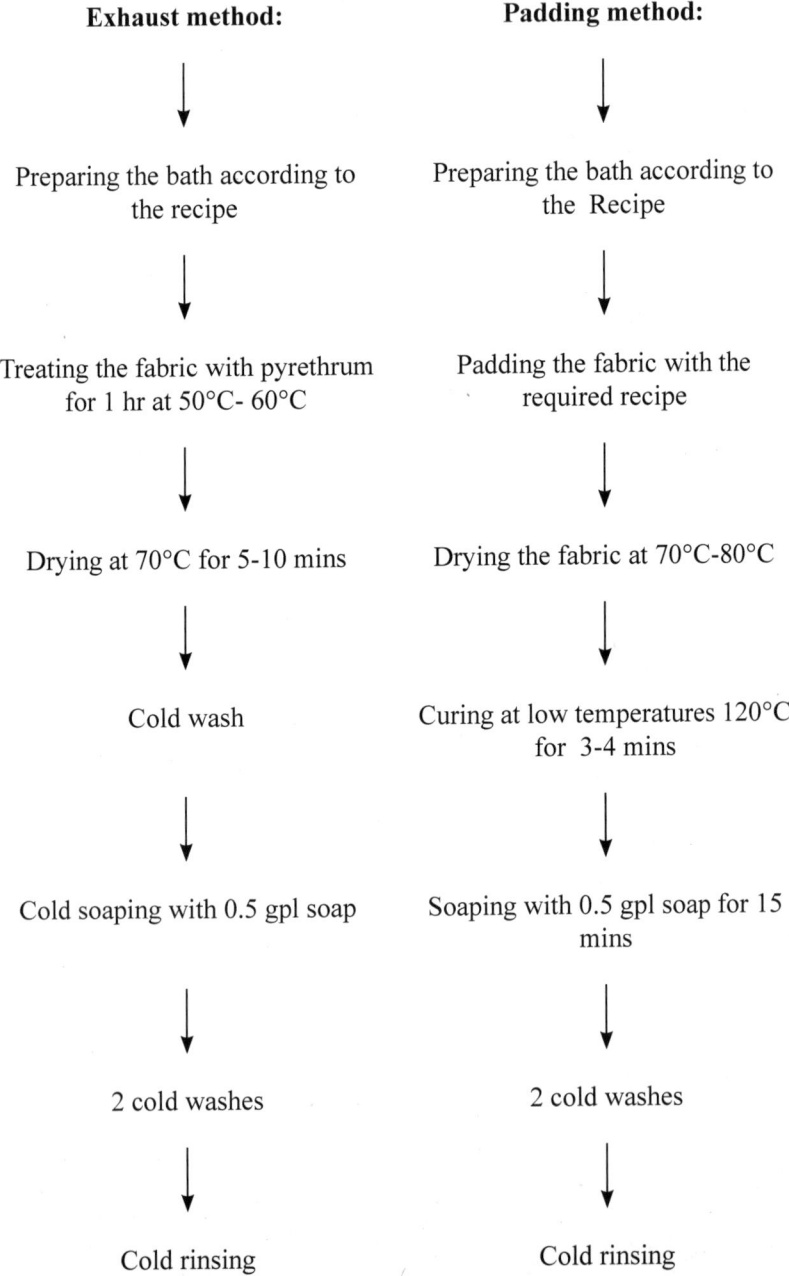

Exhaust method:

↓

Preparing the bath according to the recipe

↓

Treating the fabric with pyrethrum for 1 hr at 50°C- 60°C

↓

Drying at 70°C for 5-10 mins

↓

Cold wash

↓

Cold soaping with 0.5 gpl soap

↓

2 cold washes

↓

Cold rinsing

Padding method:

↓

Preparing the bath according to the Recipe

↓

Padding the fabric with the required recipe

↓

Drying the fabric at 70°C-80°C

↓

Curing at low temperatures 120°C for 3-4 mins

↓

Soaping with 0.5 gpl soap for 15 mins

↓

2 cold washes

↓

Cold rinsing

10.4.9   Test methods

10.4.9.1   Cone bio cone test

In this test method, the given sample is fixed at the base of the hollow cone. The cone is made out of pure plastic and has a base diameter of 11.4cms and a height of 5.7 cms. The nose part of the cone is provided with a small hole through which the mosquito species are sent in by means of a suction tube. After the mosquitoes have entered the cone, the nose part is sealed by placing a bunch of cotton. Care should be taken to provide adequate air circulation for the mosquitoes. Otherwise, it may die due to suffocation and the final results might not be conclusive.

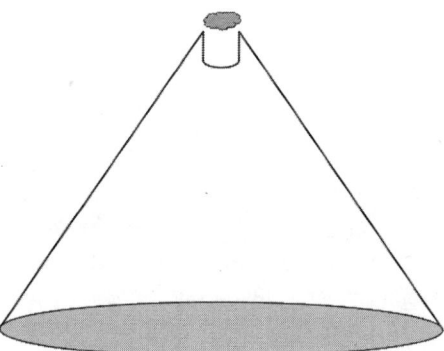

Figure 10.1: Cone bio cone

The test is carried out by taking two such cones one is called control specimen and the other one is called test specimen. In the control specimen, the untreated fabric is fixed to the bottom of the cone and in the test specimen; the treated fabric is fixed to the bottom of the cone. Both the specimens are then fixed to the wall with the base facing the wall. The behavior of the mosquitoes inside the cone is observed for a certain time period and reported.

10.4.9.2   Susceptibility test

In this test method, the given sample is placed inside the cylindrical compartment. The upper and the lower part of the cylinder are sealed with the help of a plastic lid. The lid is provided with mesh clothing in the centre to provide necessary air circulation. In this test, also two susceptibility test tubes are taken. One tube acts as the control (untreated fabric) and the other acts as the test. (treated fabric). The bottom lid has a sliding door with a small hole. The mosquitoes are collected by means of a suction device and are introduced into the control as well as the test specimen. After the mosquitoes are exposed to the treated and untreated fabrics their behavior over a time period is studied and reported.

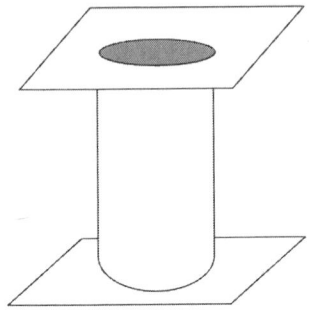

Figure 10.2: Susceptibility tube

10.4.9.3   Barad cage

In this test method, the given sample is cut according to the dimensions of the cage. In this method also, a control and a test specimen are taken for testing. Inside the cage, five fabrics are placed one on each side, the remaining one side of the cage is left free. A small provision is made at one side of the cage through which the mosquitoes are introduced. It's a usual practice that this side of the cage is kept free from the test or control specimen. The mosquitoes are left inside the cage and they are observed for their behavior for a given time period.

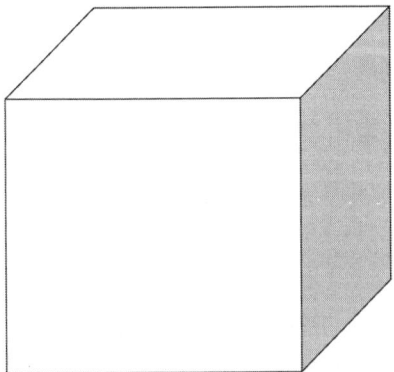

Figure 10.3: Barad cage

10.5   Results and discussion

In this project, a detailed analysis of the efficiency of pyrethrum treated fabrics has been made. All the fabrics were tested by three different test methods namely susceptibility test, cone bio cone test and Barad cage test.

For each test, ten Aedesaegyptia mosquitoes were taken and studied for their normal behavior, abnormal behavior and death at room temperature for a time period of about 1 hour. The results are tabulated in Tables 10.1.and 10.2 and their corresponding Figures are from Fig 10.1. to Fig 10.4.

10.5.1   Cotton fabrics

From Table 10.1, it is clear that at higher concentrations the number of mosquitoes died is more. Table 10.1 also shows that apart from the concentration of pyrethrum the space availability inside the testing equipment and the time of exposure of the treated fabric inside the testing equipment are also the important factors influencing mortality.

Barad cage is more spacious compared to susceptibility tube and cone. Hence the number of mosquitoes died in susceptibility tube and cage are more and these two tests are more efficient compared to Barad cage.

In Fig 10.4, the plot of concentration of pyrethrum against mortality rate decreases. Therefore, it is evident that at higher concentrations the mortality rate decreases.

According to WHO standards all the tests should be carried out for a time period of at least 24 hours. When all the pyrethrum treated cotton fabrics are exposed for 24 hours 100% mortality is observed in all the test methods.

Even though it is stated that higher concentrations of pyrethrum lead to a higher mortality rate, the ANOVA results show that there is no much significant difference in mortality rate due to concentration.

Since there is no significant difference in mortality rate due to concentration and 100% mortality is observed in all the tests samples when exposed for 24 hours, cotton fabrics with 5% and 10% pyrethrum concentrations can be chosen. 10% concentration is also stronger in all the cases.

So considering all the above-mentioned factors and economical limitations into account it is suggested to choose a cotton fabric with 5% pyrethrum concentration as an optimum concentration. In some cases, cotton fabrics with concentrations of less than 5% is also effective in their repellent action.

Table 10.1: Mosquito repellent tests on cotton fabrics treated with pyrethrum of different concentrations under different test methods.

Conc of pyrethrum	1%			5%			10%			25%		
Tests	Susceptibiliy	Cone-Bio-Cone	Barad Cage	Susceptibility	Cone-Bio-Cone	Barad Cage	Susceptibility	Cone-Bio-Cone	Barad Cage	Susceptibility	Cone-Bio-Cone	Barad Cage
No of mosquitoes	10	10	10	10	10	10	10	10	10	10	10	10
Mosquitoes with normal behaviour	3	4	5	3	3	5	2	2	3	1	2	2
Mosquitoes With Abnormal Behaviour	6	5	5	5	5	5	5	4	6	5	5	6
Mosquitoes died	1	1	0	2	2	0	3	2	1	4	3	2

Table 10.2: Mosquito repellent tests on nylon fabrics treated with pyrethrum of different concentrations under different test methods.

Conc of pyrethrum	5%			10%			25%		
Tests	Susceptibility	Cone-bio-cone	Barad cage	Susceptibility	Cone-bio-cone	Barad cage	Susceptibility	Cone-bio-cone	Barad cage
No of mosquitoes	10	10	10	10	10	10	10	10	10
Mosquitoes with normal behaviour	3	3	4	2	2	3	1	2	3
Mosquitoes with abnormal behaviour	5	6	6	5	6	6	5	5	5
Mosquitoes died	2	1	0	3	2	1	4	3	2

10.5.2   Nylon fabrics

From Table 10.2., it is clear that at higher concentrations the no of mosquitoes died is more. Table 10.2 also shows that apart from the concentration of pyrethrum the space availability inside the testing equipment and the time of exposure of the treated fabric inside the testing equipment are also the important factors influencing mortality.

Barad cage is more spacious compared to susceptibility tube and cone. Hence the no of mosquitoes died in susceptibility tube and cage are more and these two tests are more efficient compared to Barad cage.

In Fig 10.4 the plot of concentration of pyrethrum against mortality rate decreases. So it is evident that at higher concentrations the mortality rate decreases.

ANOVA: Susceptibility test

			Mortality rate											
	5%					10%					25%			
Cotton	2	0	1	0	1	3	2	2	1	2	3	4	2	4	3
Nylon	1	1	0	2	1	2	1	2	2	1	2	1	1	3	2

Row Total = 53

Column Total = 53

H_{00} – There is no significant difference between substrates

H_{01} – There is significant difference between substrates

H_{10} – There is no significant difference in mortality rates due to concentration

H_{11} – There is significant difference in mortality rates due to concentration

Source of variation	Sum of squares	Degrees of freedom	Mean sum of squares	F_{cal}	F_{tab}
Between columns	10.37	14	0.741	0.634	2.53
Between rows	1.64	1	1.64	1.403	4.60
Residual	16.36	14	1.169	–	–
Total	28.37	29	–	–	–

Since $F_{cal} < F_{tab}$ there is no significant difference between pyrethrum treated cotton and nylon fabrics and there is no significant difference in mortality due to different concentrations of pyrethrum.

Cone bio cone test:

	Mortality rate													
	5%					10%					25%			
Cotton	0	1	0	1	2	2	2	1	2	1	2	3	2	4	2
Nylon	0	1	1	1	2	1	0	2	1	1	1	2	2	1	2

Row Total = 43

Column Total = 43

H_{00} – There is no significant difference between substrates

H_{01} – There is significant difference between substrates

H_{10} – There is no significant difference in mortality rates due to concentration

H_{11} – There is significant difference in mortality rates due to concentration

Source of variation	Sum of squares	Degrees of freedom	Mean sum of squares	F_{cal}	F_{tab}
Between columns	13.87	14	0.991	1.77	2.53
Between rows	1.64	1	1.64	2.92	4.60
Residual	7.86	14	0.561	–	–
Total	28.37	29	–	–	–

Since $F_{cal} < F_{tab}$ there is no significant difference between pyrethrum treated cotton and nylon fabrics and there is no significant difference in mortality due to different concentrations of pyrethrum.

Barad cage:

	Mortality rate													
	5%					10%					25%			
Cotton	0	1	0	2	1	2	1	2	3	0	1	3	2	2	3
Nylon	0	1	0	1	1	0	1	1	2	1	1	1	2	1	1

Row Total = 37

Column Total = 37

H_{00} – There is no significant difference between substrates

H_{01} – There is significant difference between substrates

H_{10} – There is no significant difference in mortality rates due to concentration

H_{11} – There is significant difference in mortality rates due to concentration

Source of variation	Sum of squares	Degrees of freedom	Mean sum of squares	F_{cal}	F_{tab}
Between columns	8.87	14	0.634	0.753	2.53
Between rows	2.71	1	2.71	3.219	4.60
Residual	11.79	14	0.842	-	-
Total	23.37	29	-	-	-

Since $F_{cal} < F_{tab}$ there is no significant difference between pyrethrum treated cotton and nylon fabrics and there is no significant difference in mortality due to different concentrations of pyrethrum.

10.5   Conclusion

In this world, where most of the dreadful diseases are spread by mosquitoes many alternatives are provided to encounter this problem. However, every alternative has some drawbacks one way or the other. In this project, an attempt has been made to provide a safe alternative and it was concluded that fabrics treated with pyrethrum are very effective against mosquitoes.

In this work, various tests are carried out with pyrethrum treated fabric samples against the species such as *aedes, anopheles,* and *varigillus* species. The test results were very conclusive as 100% mortality rates were observed with all the species. Also, pyrethrum treated fabrics are found to be safe for humans.

The following conclusions have arrived from this project:

1. Good mosquito repellency was observed in both cotton and nylon fabrics treated with pyrethrum.

2. 100% mortality rate was observed in all the pyrethrum treated samples when the mosquitoes were exposed for 24 hours.

3. Fabrics with pyrethrum concentration of 5% were found to be optimum.

4. Susceptibility tube and cone bio cone gave better results when compared to Barad cage.

5. For fine nylon mesh cloths, exhaust method or batch process is suitable.

Both cotton and nylon fabrics treated with pyrethrum, posseses good wash fastness properties (finish remains even after 5-10 washes).

10.6   Reference

1. Clements AN. The Physiology of Mosquitoes. Oxford: PergamonPr; 1963.

2. Maibach HI, Skinner WA, Strauss WG, Khan AA. Factors that attract and repel mosquitoes in human skin. JAMA. 1966; 196:263-6.

3. Bock GR, Cardew G, eds. Olfaction in Mosquito-Host Interactions. New York: J Wiley; 1996.

4. Gjullin CM. Effect of clothing color on the rate of attack of Aedes mosquitoes. J Econ Entomol. 1947; 40:326-7.

5. Gillies MT. The role of carbon dioxide in host-finding by mosquitoes (Diptera: Culicidae): a review. Bulletin of Entomological Research. 1980; 70:525-32.

6. Davis EE, Sokolove PG. Lactic acid-sensitive receptors on the antennae of the mosquito, Aedesaegypti. J Comp Physiol. 1976; 105:43-54.

7. Khan AA. Mosquito attractants and repellents. In: Shorey HH, McKelvey JJ, eds. Chemical Control of Insect Behavior. New York: J Wiley; 1977:305-25.

8. de Jong R, Knols BG. Selection of biting sites by mosquitoes. In: Bock GR, Cardew G, eds. Olfaction in Mosquito-Host Interactions. New York: J Wiley; 1996:89-108.

9. Kline DL, Schreck CE. Personal protection afforded by controlled-release topical repellents and permethrin-treated clothing against natural populations of Aedestaeniorhynchus. J Am Mosq Control Assoc. 1989; 5:77-80.

10. Schreck CE, Kline DL, Carlson DA.Mosquito attraction to substances from the skin of different humans. J Am Mosq Control Assoc. 1990; 6:406-10.

11. Knols BG, de Jong R, Takken W. Trapping system for testing olfactory responses of the malarial mosquito Anopheles gambiae in a wind tunnel. Med Vet Entomol. 1994; 8:386-8.

12. Geier M, Sass H, Boeckh J. A search for components in human body odour that attract females of Aedesaegypti. In: Bock GR, Cardew G, eds. Olfaction in Mosquito-Host Interactions. New York: J Wiley; 1996:132-48.

13. Foster WA, Hancock RG. Nectar-related olfactory and visual attractants for mosquitoes. J Am Mosq Control Assoc. 1994; 10 (2 Pt 2):288-96.

14. Curtis CF, Lines JD, Ijumba J, Callaghan A, Hill N, Karimzad MA. The relative efficacy of repellents against mosquito vectors of disease. Med Vet Entomol. 1987; 1:109-19.

15. Muirhead-Thomson RC. The distribution of anopheline mosquito bites among different age groups. Br Med J. 1951; 1:1114-7.

16. Gilbert IH, Gouck HK, Smith N. Attractiveness of men and women to Aedesaegypti and relative protection time obtained with DEET. Florida Entomologist. 1966; 49:53-66.

17. Port GR, Boreham PFL. The relationship of host size to feeding by mosquitoes of the Anopheles gambiae Giles complex (Diptera: Culicidae). Bulletin of Entomological Research. 1980; 70:133-44.

11

Garment processing

Dr. P. Senthilkumar

Professor, Department of Textile Technology, PSG College of Technology, Coimbatore - 641004, Email: senthiltxt11@gmail.com

Abstract : The changes affected in the retailing sector with respect to product mix and fashion trends has evoked a response in the textile manufacturing industry for increased dyeing and finishing of fully fashioned garments. The recent surge in export of garments has created a challenging opportunity for the garment processing industry because of the short pipeline necessary for quick response. During garment processing, there are reduced losses from dyed fabric waste. Presently, interest in garment processing is very high and there is a scope in it. Garment processing is in no way different in principles and in technical sense to fabric dyeing or processing. It should not be compared as a substitute or an alternative to fabric dyeing but should be considered complimentary to a wet processing unit. The popularity of a garment dyeing seems to have started in the denim field. At present garment processing, true dyeing of garments and special finishing effects on garments are undergoing a rapid growth.

Keywords : Garment processing, garment dyeing, garment finishing, garment dyeing machines

11.1 Introduction

At present, interest in garment dyeing is very high and there is a scope in it, acid washed jeans or stone washed garments represent the major volume of garment production at present. The dyeing of garments is undergoing rapid growth and, in fact, the demand is currently greater than the available capacity. The lack of technical personnel in this segment is another contributory factor for this industry not being able to cope up with the present day requirements of finished items of garment exports.

Garment dyeing/processing is in no way different in principles and in a technical sense to fabric dyeing or processing. It should not be compared as a substitute or an alternative to fabric dyeing but should be considered complimentary to a wet processing unit.

The popularity of a garment dyeing seems to have started in the denim field, wherein stiff, boardy jeans gave way to prewashed jeans next followed the stone washed jeans. The latest fad is a spotty, bleached effect, which goes by a number of different names, i.e., acid washed, ice washed, whitewashed, snow-washed, etc. Although ice washed, stonewashed or sand washed jeans and dress material represent the major volume of garment processing at present, true dyeing of garments and special finishing effects on garments are undergoing a rapid growth [1].

The changes affected in the retailing sector with respect to product mix and fashion trends have evoked a response in the textile manufacturing industry for increased dyeing and finishing of fully fashioned garments. The recent surge in the export of garments has created a challenging opportunity for the garment processing industry because of the short pipeline necessary for a quick response. During garment processing, there are reduced losses from dyed fabric waste. Most importantly to produce uniformly dyed garments with no variations in the shade of the different parts of the garment, which is a major problem for manufacturers using piece-dyed fabric because of sort-to-sort variations [2].

There are both advantages and disadvantages inherent in garment dyeing, but looking at the present industrial conditions of textile industry of India, there are more advantages in garment dyeing/processing in comparison to fabric processing, though about 90% of garments dyed and processed at present in the garment form are cut from previously prepared cloth. It is, therefore, to be considered Garment dyeing/processing not as a substitute to fabric dyeing but as a complimentary to fabric dyeing. Garment manufacturers should also select garment-processing procedures suited to these fabrics and garments, which are amiable and economical as compared to fabric processing [3].

11.1.1   Advantages of garment dyeing/processing

- All loosely woven or knitted or hosiery fabrics are best dyed or processed by garment processing.
- Handloom fabrics loosely woven or bigger width bedcovers etc can be dyed and processed in garment dyeing machines.
- Jeans which are warp dyed can be better with less energy consumed, such as avoiding calendaring and double preshrinking on sanforizing ranges.
- Striped or overdyed material like shirting or suiting material can best be dyed in the garment form economically and for the saving of inventory and leftover of unstitched material.
- Stonewash – acid wash – crepe- georgette material and seersucker material are best processed in the garment form, both for the fashion purpose as well as for keeping the shape of the garment intact several washing.

- To provide longer lengths and bigger widths of processed cloth for export, it is almost impossible with the installed machinery in the conventional process houses. If garment processing procedures are followed it can cater to the needs for value added exports of textile materials.
- Permanent press, water repellent, and fire retardant garments can be exported easily through garment processing procedures.
- Research is needed in the sizing section to produce a sizing recipe, which is washed out easily. In that case, the grey cloth can be directly taken for garment dyeing procedures, irrespective of the construction of the cloth.
- In fact, it is suggested that it is better to concentrate on producing fabrics for export which do not require sizing at all and which can be taken out for dyeing and subsequent processing in the garment form only.
- Garment dyeing/processing is not a substitute or alternative to fabric dyeing but garment processing is definitely complementary to fabric processing and garment stitchers or readymade manufacturers. It is advisable to concentrate only in the field, which is not being sold or exported in the fabric form at present for value added exports. There is a dearth of qualified technicians and scientists in the laundry filed. There is a lot of scope in garment processing especially to:
  - cater to the needs of small scale readymade manufacturers.
  - develop new avenues of processing for value added exports.
  - cater to the newly developing needs of hotel industry or laundry industry.
  - cater to the needs of the terry towel industry at economical rates.
  - cater to the needs of the hosiery industry both for natural and manmade knitted fabrics.
- To modernize the wet processing department, for taking care of bigger width cloths in longer lengths is not easy or economical. Investment in garment dyeing/processing will be financially viable.
- All technicians or even manpower required from wet processing can be usefully utilized from existing staff.

11.1.2   Disadvantages of garment dyeing/processing

- Random selection of garments without considering the limitations of 3M`s i.e. Men, Machines and Material.

- Many examples are found wherein garments are stitched from mixed fabrics, such as bleached and unbleached cotton, mercerised and unmercerised cotton sewn together, and pockets of material different and even resin treated cloth sewn to greige goods.

- A given garment must be cut from one piece or a roll of fabric and that excellent preparation is required prior to dyeing.

- Inadequate preparation: It is highly important that the preparation of cloth be fully insured before undertaking dyeing. If it is not satisfactory then there is a great danger that dyeing will be faulty and if the garment is stitched with fabric from various places, then different components will dye differently. Cases are known with dye to pick up different at pockets collar and sleeve obviously due to cloth from different lots.

- Just as with conventional dyeing, it is estimated that 70% of all dyeing problems are due to improperly prepared material. It is thus highly important that garment dyeing receive more than full standard preparation.

- It is a batch process and thus the cost of production will be comparatively more than the conventional high capacity or continuation machines used for fabric dyeing.

- Problems other than preparation can also be encountered in garment dyeing due to the construction of the garment. While they are not necessary for the preview of the garment dyer, because they are the factors which influence the acceptability of the dyed garments.

- Problems of shrinkage in a garment: It is important that the fabric should be consistent from one source. In India, however, it is difficult to get consistently the fabric of the same construction or even semi-processed cloth from the same source unless it is obtained from one composite mill. Under these circumstances, it is said that a company's success in cutting all parts of garments from the same roll of cloth, accurately establishing shrinkage (prior experiments) will develop garments to a satisfactory level. In other words, there is an importance of having a well-trained capable dyer and up to date dye house, which unfortunately is somewhat difficult as technical personnel and skilled labour for garment processing are a scarce commodity now.

11.1.3　Structure of the garment-dyeing sector

1. Conventional finishers - Traditional finishing plants that have adopted garment dyeing in addition to their conventional production.

2. Laundries/Drycleaners – These establishments do garment dyeing as an additional profitable activity, to give a new lease of life to garments with no guarantees offered on the final fastness properties.

3. Garment dyers – Composed of established fully fashioned dyers.

11.2   Quality control in garment processing

The garments produced from woven fabrics have created many problems, and experience has shown that existing styles as developed for piece-dyed fabric cannot be just assembled from grey fabric and thrown into the dyeing machine. Unless they have engineered it from the original design stage for garment dyeing. During garment processing, care should be taken that the grey fabrics dye-up identical in shade and if necessary must be adequately pretreated, before dyeing [4].

The other major areas for control are:

• Seams, elasticated areas, waistbands, cuffs.

• Shrink behavior.

• Chafe marks/creases.

• Accessories

• Sewing thread

• Foreign substances

• Interlinings, care labelling depending on the fastness requirement, being the most important of them all.

Elasticated areas, waistbands, and cuffs must be slack, and seams should not be prepared too tightly or bulky or poor penetration of dyestuff occurs especially with heavy articles and heavily swelling fibres like cotton. However, the problem can be solved by using suitable dyes and the right process control.

The use of a high application temperature dyes not only ensures that the migration potential is fully realised, it also offers several additional advantages. Higher temperature means better diffusion, better penetration, and better running of the cloth facilitating liquor flow. These benefits assume particular significance in garment dyeing, especially in garments that have multilayered seams of woven fabrics (pockets, lapels, zips, etc.) or tight elasticated waists and ribs.

Thus, a short period at high temperature i.e., 95°C before cooling for fixation with reactive dyes has overcome even the most difficult seam penetration problems.

11.2.1   Shrink behaviour

Shrink behaviour is important because excessive shrinkage of the article or uneven shrinkage, where knitted and woven fabric are mixed lead to seam puckering when it becomes important to pre-relax knitted fabric and the pretreated woven fabric must be fully preshrunk by giving the various shrink proofing treatment for high quality fabrics. To rule out the possibility of the later complaint it is advisable to carry out blank dyeing.

11.2.2   Chafe marks/creases

Chafe marks and creases are a problem particularly for delicate articles due to the mechanical stress in drum dyeing machines which are prone to chafe marks or piling effects should be turned inside out and dyed with the addition of a non- foaming lubricant while adopting the shortest possible dyeing process. With woven fabrics, certain types of weaves are liable to creasing and breaks. One preventive measure from the outset is to avoid overloading the dyeing drum.

11.2.3   Accessories

Care must be taken with the selection of zips and material accessories, such as buttons and studs. Ferrous metal must be avoided, and good quality components fabricated from nickel or its alloys must be used if breakdown through corrosion in bleaching or reactive dyeing, with its high concentration of electrolyte and alkali, is not to be a problem.

The composition of buttons can be a difficult choice. Some buttons, such as those of casein and cellulose, break up under the dyeing conditions, while polyester buttons do not dye at all and must be used either as a neutral colour, capable of coordinating with a range of shades or be purchased ex-stock. Nylon buttons can be coloured in the garment process but it is a difficult procedure to control, to obtain consistent results.

Elastomerics, either natural rubber or polyureathane fibres such as Lycra, are used in ribs and both types can give rise to problems. Natural rubber is adversely affected by certain metals, and for this reason, it is important to use dyestuffs that do not contain copper. Polyurethane fibres are seriously weakened in the presence of strong oxidizing agents such as chlorine, but hydrogen peroxide bleaching can be used with the polyester type of elastomeric fibre. Hence, it is very important to specify the correct Lycra tape to avoid a breakdown in dyeing. To prevent the major problems due to the metal accessories corrosion protection agents must be used.

11.2.4   Sewing thread

Sewing thread is undeniably one of the most vital components to be found in countless articles, which are used, in day-to-day life. It is a component taken for granted and the demands on the threads are so many that it must satisfy a wide range of needs [5].

Properties desirable for sewing threads:

1. It must be strong and yet fine enough to produce a neat seam and last for the whole life of the product.
2. Considering the modern high-speed sewing machines the thread must be designed to withstand the speeds. An additional lubricant is required to reduce friction between the thread, the needle, and the thread, the needle, and thread guides and the fabric and protect synthetic-fibre threads against needle heating. Therefore, before dyeing these lubricants have to be removed, since otherwise, it will inhibit dye uptake by the sewing thread.

11.2.4.1   Selection of fibre type for the thread

For garment dyeing the thread and fabric have to be of the same fibre type, however, this is suitable mainly for cotton, as sewing threads are not produced in all fibre-types because of the dyestuff specific nature.

11.2.4.2   Thread selection and precautions in selection

In sewing fabrics made from cotton in order to obtain equivalent seam strengths, heavier cotton threads must be used. Further, it is necessary to keep sewing tensions to a minimum on lockstitch and chain stitch operation deliberately producing slack stitches since otherwise pucker will be introduced owing to shrinkage of the thread during subsequent garment dyeing.

All these factors would slow down the production speeds for garments to be post dyed and hence costing should be carefully adjusted. Whichever thread is selected ultimately it will be a trial and error situation and hence sampling is essential before bulk dyeing.

11.2.5   Foreign substances

To reduce value loss, it is better to prevent foreign substances such as stains due to oil, grease and other lubricating agents used during the course of manufacturing. Further proper handling and good housekeeping also would reduce the staining.

Other foreign substances could be sizes and resins containing additives such as elastomers and oil repelling agents, which would cause poor appearance of the final goods and hence must be taken care of before dyeing.

11.2.6   Interlining

The purpose of an interlining, whether fusible or non-fusible, is to stabilize and add body to the outer fabric. With the introduction of post-dyed garments, certain special properties had to be introduced into interlinings to ensure that they performed satisfactorily.

The special properties required are:
- It must take up dye from the bath to a similar level as that of the outer fabric.
- The adhesive, bonding the outer fabric to the interlining base must remain intact during and after the dyeing operation.
- The handle after dyeing should be that required by the customer.
- There should be no adverse colouration caused by dye absorption in the fusible resin used in the fusible. The base fabric, base-fabric finish and the fusible resin coating three main components of any fusible, which should have the above properties.

The best choice for the base fabric is cotton, as it would not cause problems during dyeing. Resin could be used however, they take up the normal hydroxyl sites needed for dyeing and hence often a combination of a low formaldehyde resin and mechanical shrinking has to be done to get even dye absorption and prevent the differential shrinkage. The fusible resins must be resistant to breakdown during dyeing and hence they should have the following properties such as the resins should be hydrophobic, inert and have high melting points.

11.3   Dyeing procedures

At present, the only garments being dyed commercially are 100% cotton although it is felt that blended fabrics would be seen. The dyes used are direct, fiber reactive, vat and sulphur. The usual principles of dye selection and applications apply to garment dyeing, special care should be taken to select those dyes and methods which give the best levelling and penetration.

All the general and well-known practices for applying a dye to cotton should be followed so that satisfactory penetration is achieved as well as good colour uniformity through the batch. To prevent abrasion problems, sensitive garments should be turned inside out prior to dyeing. It is generally felt that loading of the dyeing machines should not be more than 50 to 66% of the rated capacity.

11.4   Garment dyeing machines

A great variety of machines for garment dyeing is available in the market, which exhibits widely different designs. Since it is virtually impossible to present a complete survey, this section will present a summary of the different types of construction, their main features, and their method of operation. These types of garment dyeing machines can be divided as follows [5]:

1. By the manner in which the goods and the liquor are moved
    (a) Mechanically by a paddle
    (b) Mechanically in a rotating drum
    (c) Hydro-dynamically by adjustable jets
    (d) Hydro-dynamic circulating dyeing machines with Floating Liquor Circulation principle

2. By the type of construction
    (a) Padding machines (horizontal, oval & High Temperature-HT paddle types)
    (b) Drum machines
    (c) Washing-centrifuging machines
    (d) Drum dyeing centrifuging machines
    (e) Jet dyeing centrifuging machines
    (f) Hydrodynamic circulation dyeing machines with Floating Liquor Circulation principle.

11.4.1   Paddle machines

Three types of paddle machines are used:
- Horizontal paddle machine
- Lateral/oval paddle machine
- HT paddle machine

11.4.1.1   Horizontal paddle machine (overhead paddle machine)

This machine consists of a curved beck-like lower section, which contains the material and the dye liquor. The goods are moved by a rotating paddle, which extends over the width of the machine. The paddle is just about half-submerged in the liquor and moves the dye liquor evenly over the entire width of the back.

The design of the paddle causes the material to move upwards and downwards through the liquor. The material and the liquor move in axial circular movement (i.e., from the front of the machine in a circular motion downwards and towards the rear). Dyeing can be carried out at temperatures up to 98°C.

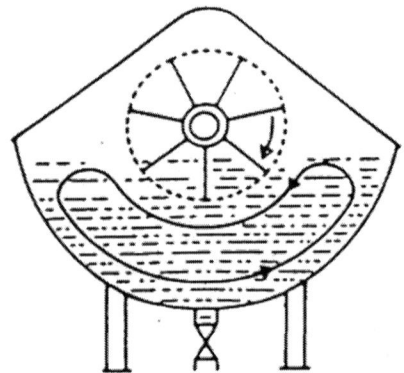

Figure 11.1: Horizontal paddle machine

11.4.1.2   *Lateral/Oval-paddle machine*

To avoid any point in the bath with poor liquor flow, the machine consists of an oval tank. In the middle of this tank is a closed oval island around which the liquor is kept in motion. The paddle moves in a lateral direction and is not quite half-submerged in the liquor. Here temperatures are set around 98°C.

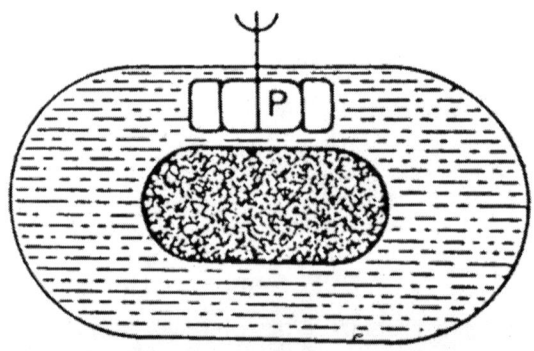

Figure 11.2: Lateral/oval paddle machine

11.4.1.3   HT paddle machine

This machine operates according to the horizontal paddle principle. The liquor receptacle is cylindrical and designed so that dyeing can take place at a temperature of up to ca. 140°C.

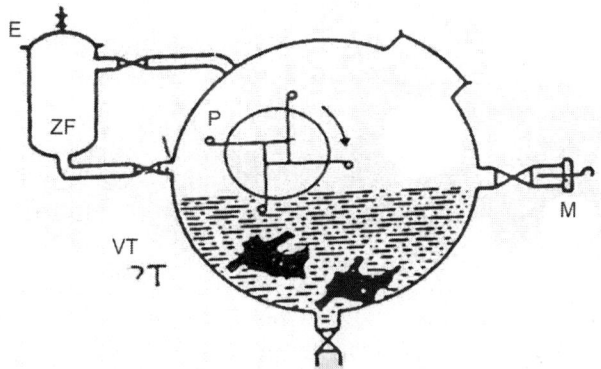

Figure 11.3: HT paddle machine

11.4.1.4   Dyeing with paddle machines

Liquor ratio

Dyeing is carried out at liquid ratios of 30:1 to 40:1. Lower ratios than this are to be avoided, as this would entail a reduction of optimum movement of the goods. This leads to unlevel dyeings and heavy creasing, with a considerable amount of work.

Dyeing

For gentleness to the goods, the blades of the paddle are either curved or have rounded lengthwise edges. The rotating speed of the paddle can be regulated from 1.5 to 40 rpm., depending on the quality of the goods and the final appearance desired. In most cases, a speed of 10 to 20 rpm is used.

The circulation of the dye liquor should be just strong enough to prevent the goods from sinking to the bottom. The goods are normally dyed loosely packed in polyester or polypropylene bags (up to 70%of their capacity).

According to the machine, article dyed and desired results. The goods can also be dyed unpacked. Where high quality requirements must be met the goods are turned inside out. Again, overloading the machine should be avoided so as not to risk reduced movement of goods with resulting high likelihood of unlevel dyeing. The loading capacity of paddle machines ranges from 1 to 25 kg of dry weight.

Paddle machines are suitable for dyeing articles of all substrates in all forms of make-up. Polyester articles are preferably dyed on HT paddle machine and Acrylic articles on the same type of machine at 105 to 107°C.

11.4.2   Drum dyeing machine

In this machine, a perforated drum is suspended along a lengthwise axis in a horizontal position, submerged in the dye liquor. To ensure good movement of the goods, the drum is divided into four equally large compartments separated by crosswire dividing walls. These four compartments can be loaded and closed individually. The drum is capable of rotating in either direction (including alternating directions). The rotating speed is steplessly adjustable from 2 to 20 rpm.

Besides classic drum machines for dyeing at 98°C, there are also HT drum machines for dyeing at room temperatures up to 140°C. The speed and direction of the rotating drum are decisively important for the results. Normally, low rotating speeds are used. The degree of loading and whether or not the goods are dyed in bags are also factors which determine the final appearance of the dyed goods. Centrifuging is not possible on conventional drum machines.

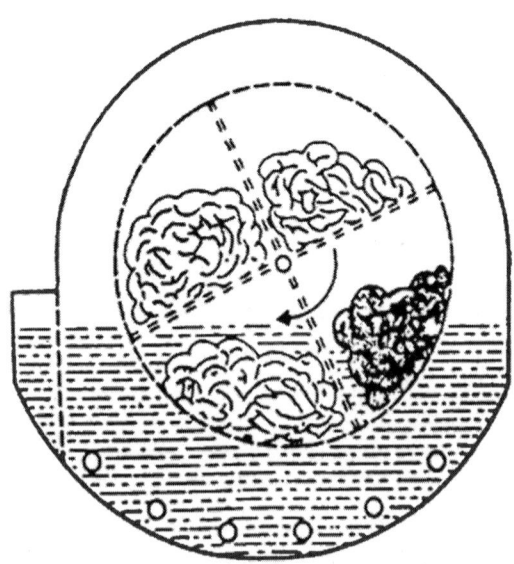

Figure 11.4: Drum dyeing machine

11.4.3   Washing-centrifuging machine

This type of machine is similar in construction to a normal household washing machine. The washing-centrifuging machine has a suspended perforated inner drum which is not divided into compartments. The inner drum is located within an outer drum. The washing-centrifuging machine differs only in technical respects from household washing machines, i.e., according to its purpose, it is equipped with a feed tank, sampling port, and variable speed and dyeing programmes selection. The further industrial development of this machine has led to the drum centrifuging machine.

11.4.4   Drum dyeing-centrifuging machine

This machine is also called a multipurpose drum machine or multi-rapid dyeing-centrifuging machine, since it can perform all wet processing operations such as scouring, dyeing, conditioning and centrifuging successively with automated program control. The machine is distinguished by the fact that it permits rapid loading and unloading, can operate with a very low liquor ratio and is available for dyeing at temperatures up to 98°C or as HT machine for temperatures up to 140°C.

The interior of the machine looks like that of a household washing machine where the liquor and the goods are agitated by the rotating drum. Machines are also available where the liquor circulation through the goods can be intensified by building in additional jets.

The goods are treated in a perforated inner drum housed within an outer drum or dyeing tank. The inner drum can be supported at one or both sides and divided into compartments or not. Inner drums without dividing walls are provided with ribs that carry the goods along for a certain time, partially lifting them up out of liquor. The inner drum may also be divided into 2 or 3 compartments with perforated or solid dividing walls. The Y design has proved to be the most advantageous form of drum partitioning, where perforated dividing walls serve to increase liquor circulation yet further. The drum rotating speed can be infinitely varied and set in one or alternating directions.

The goods can be dyed loose or packed in bags. Depending on the article and quality requirements after the wet treatment is completed it is possible to centrifuge the goods to a point where there is no need for a separate tumble drying operation.

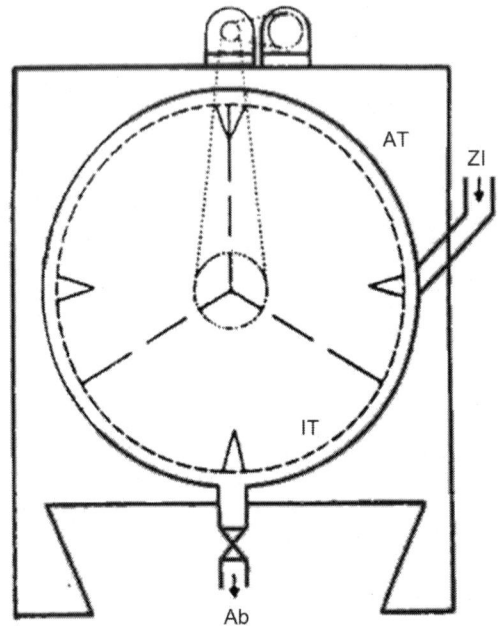

Figure 11.5: Drum dyeing centrifuging machine

11.4.5 Drum dyeing-centrifuging machine

Liquor ratio

Besides 4:1 to 18:1 liquor ratios, machines are also available which allow stepless adjustment of liquor ratio. These offer the advantage of retaining the liquor ratio which corresponds to the amount of material dyed thus producing good reproducibility with dyeings of varying batch size.

Besides dyeing-centrifuging machines with only one drum, models have been constructed with two drums alongside one another. This permits doubling the size of the batches without the disadvantages of one drum with an oversized diameter.

11.4.6 Jet circulation dyeing machines

These circulation dyeing machines are normally constructed to permit HT dyeing temperature up to 140°C. The dye liquor and the goods are kept in motion by jet nozzles whose direction and force are adjustable. Nozzles located along the wall cause the liquor to move in a circular motion.

Turbulence nozzles on the bottom ensure the liquor circulation, preventing the goods from sinking and allowing them to open. The machine load capacity ranges from 25 to 125 kg of dry weight. Jet circulation machines can be used to dye all substrates in all forms of make-up, most of them even without having to be packaged in small bags.

Liquor ratio

The liquor ratio can be varied from 25:1 to 40:1. Low ratios are to be avoided for the reasons mentioned when discussing paddle machines. For articles that tend to become entangled and are not packed in bags, the liquor ratio must be greater than 30:1.

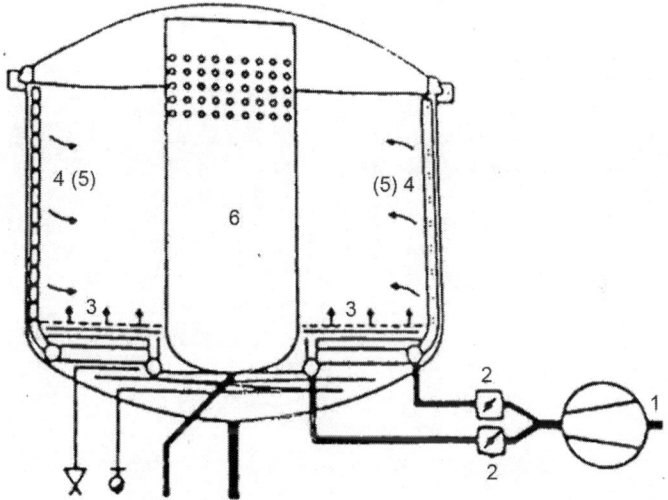

Figure 11.6: Jet circulation dyeing machine

11.4.7   Hydrodynamic circulation machine

This type of machine is also called as circulation dyeing machine with a "floating liquor circulation". It is normally sold as HT machine, allowing dyeing temperatures up to 130° C.

The diagram shows the liquor circulation principle, i.e. a pump conveys the liquor to the vertical liquor distributor (V), which rests upon a specially constructed perforated bottom(S). The liquor distributor and perforated bottom are designed so that an adjustable spiraling liquor circulation is produced which keeps the goods in movement.

The circulation can be steplessly adjusted from nil to maximum. Even with weak circulation, the goods are prevented from sinking to the bottom. The machine is suitable for all articles made of all substrates in all forms of make-up.

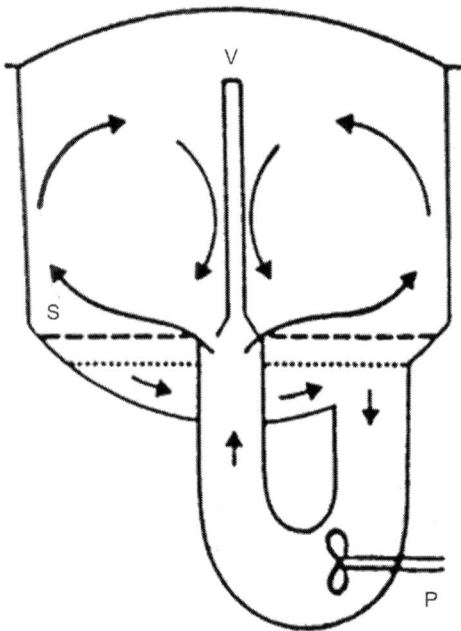

Figure 11.7: Hydrodynamic circulation machine

11.5   Machines for post-dyeing treatments

After dyeing and centrifuging, the goods must be dried and steamed and/or pressed [6].

11.5.1   Driers

Drying is carried out in a tumbler constructed of a sheet steel shell containing a perforated stainless steel drum, which rests on friction wheels. A heating unit is located at the rear of the drum, beneath which a powerful air suction unit is located. A large door on the top permits loading and unloading. Very large machines are loaded through the top and unloaded through another door at the back.

The suction unit draws in the fresh air, which passes through the heating element and is then axially suctioned through the drum and the goods. It is then drawn through a large lint sieve and out of the machine through an exhaust duct. Most tumblers come equipped with reverse drive, allowing the direction of rotation to be selected and alternated. The latest models have temperature control, thus ensuring maximum precision in the drying treatment and constant temperatures.

In the drying process, the goods are carried up by the ribbed drum to its apex and then fall of their own weight down through a current of air, which can be adjusted. This causes compression, the strength of which can be controlled by setting the air current. This process is influenced by the moisture content of the goods. Thanks to this tension-free movement in the drum, the goods can relax and thus undergo uniform relaxation shrinkage.

On modern drying tumblers the various parameters influencing the dyeing process such as temperature, moisture, steam injection, air circulation rate, drum speed, etc., can be controlled with great precision, thus ensuring reproducible drying operations.

Tumblers are also now available with a built-in steam injection for shrinking, bulking, or milling dry goods under optimum conditions.

11.5.2   Mechanical finishing treatments

Mechanical finishing treatments or simply "finishes" refer to operations which must be given to operations which must be given to made-up textiles before they are ready for delivery.

In most cases, they consist of a steaming treatment to remove creases or to relax the goods. This is carried out manually or semi-automatically [7].

11.5.3   Topper

This machine is mainly used for simultaneously steaming and stretching trousers. The trousers are fastened at the waist with clamp A and the bottom with clamp B. Plate C is then pushed downwards, stretching the trousers as desired, depending on the force applied. At the same time is blown from the waist downward (D) through the trousers.

The steaming time can be set. This treatment is similar to pressing but has the advantage of stretching the trousers, thus making it possible to prevent possible shrinkage.

Topper

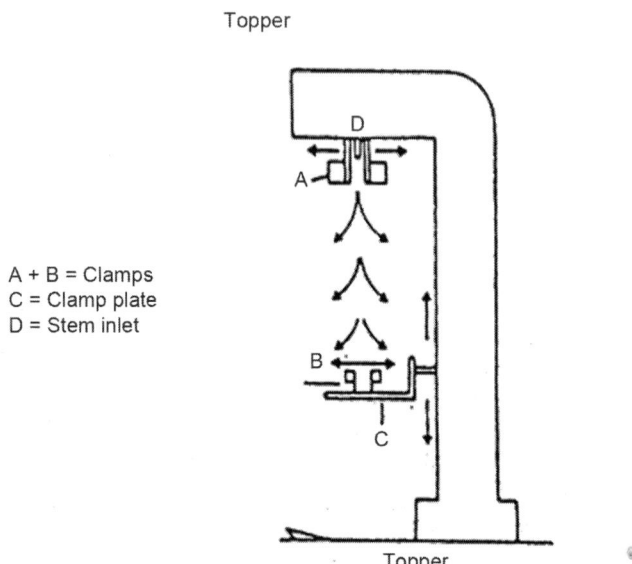

A + B = Clamps
C = Clamp plate
D = Stem inlet

Topper

Figure 11.8: Topper

11.5.4   Pressing dummy

The pressing dummy operates according to the same principles as the topper. It consists of a form over which pullovers, jackets, sweatshirts, shirts, etc., are fitted. Then steam is blown outwards through the garment. This treatment is given under relatively low tension.

11.5.5   Ironing

Ironing is the most labour-intensive mechanical finish. It is only carried out where the articles leave no other alternative. The garment is ironed by hand or machine. In most cases, ironing is done only on certain parts of the garment which are not yet free of creases after steaming on the topper or pressing dummy, e.g. collars, pockets flaps, cuffs, button flaps, etc.

11.6   Care labelling

The garment dyed-goods generally have fastness properties inferior to the normal piece dyed goods. The major properties are dimensional stability, colour fastness to washing and rubbing.

Based on the various physical and colour fastness properties a suitable care label is arrived at, which tells the customer the precautions he should take when cleaning or washing the garment. The labels are designed such that they must give specific instructions with respect to the method of washing, bleaching, drying, ironing, suitability for dry cleaning and other relevant parameters [8].

11.7   Garment finishing

After the various treatments and during the dyeing process itself, the fibre material is thoroughly degreased. Although this degreasing is necessary in order to be able to dye the goods at all, at the end of the dyeing process, they exhibit an unpleasant, hard, rough, or even brittle handle [9].

The purpose of finishing garment dyed articles is to give the goods softness and smoothness in keeping with the character of the article, with a pleasing handle and drape. Finishing can be carried out in a simple manner by the application of cationic softeners by the exhaust method at the end of dyeing process. In some cases, improvement of the elasticity and crease resistance, as well as a permanent soft handle, is also desired. Here a finish with silicone elastomers is the method of choice. The stain-resistant finishes on articles of wool and synthetics, cationic fluorocarbon products are suitable [10].

Only a little amount of garments is commercially finished and the major stress in India is only on the popularity of "distress look" in clothing. This has been one of the motivating forces, which have led to the current interest in dyeing garments. No one expects this wrinkled look to remain popular forever, so, if garment dyeing is to remain a factor in the clothing industry, an improved appearance of the garment after dyeing will be required and more so by way of maintaining the finish after several washes [11].

Garment dyeing/processing can be utilized for all types of fabrics usefully, through precautions for the selection of fabrics must be considered properly to avoid premature death in popularizing this approach. Although dyed and processed out of prepared or semi-prepared fabrics, it will not be long before grey goods will be taken up for garment dyeing/ processing as it then saves an enormous amount of energy utilized at present for fabric processing [12].

11.7.1   Finishing with softeners

The most important type of finishing for garment dyeing is the application of softeners to improve the handle with respect to softness, smoothness or bulk. Cationic softeners are mostly used. They are extremely effective and can be applied by the exhaust method without problems on all type of fibres. They are primarily suited for coloured goods.

Amphoteric softners are preferable for white goods of cellulosic fibres, wool, and polyamide, as they do not affect the whiteness produced by the usual fluorescent brighteners for these fabrics. The adsorption behaviour of cationic softners is determined by the pH and the bath temperature [11].

11.7.2   Finishing with silicone elastomers

Usage of Elastomers produces a specific surface smoothness on the textile fibres and improves the stretch and recovery properties of crease resistance of knitted goods. It confers an unsurpassed softness on the goods and facilitates the pressing and steaming of the treated articles. The effects exhibit good fastness to washing and dry cleaning [12].

11.7.3   Stain resistant finishing

A cationic fluorocarbon product can be applied on wool, synthetics, and blends by the exhaust method. It confers stain resistant properties on the fibre surface so that stains and spots do not impregnate the fabric and are easily removed.

11.8   Conclusion

Several observations about the future of garment dyeing appear reasonable. Interest in garment dyeing is high at present. Certain fabrics, such as corduroys and other low pile fabrics look better when garment dyed in garment form. Competent dyers should be utilized in garment dyeing to ensure satisfactory work. With all the advantages and disadvantages inherent in garment dyeing/processing, it will be interesting to follow this new field for the next few years, especially if technical courses on garment dyeing/processing are included early to prepare the upcoming people interested in this field.

11.9   References

1. Premal Udani, "Problems and prospects of knitwear industry", Colourage, October 1993, pp.19-22.

2. R. K. Dalmia, "Status of chemical processing for export of value-added fabrics", Colourage, April 1993, pp.23-26.

3. D .R. Sharma, "Value added exports of textile materials through garment dyeing/processing procedures", Colourage, May 1993, pp.31-35.

4. S. Y. Kamat & E.W.Menezes, "Garment dyeing – Part – I", Colourage, November 1993, pp.41-44.

5. S. Y. Kamat & E.W.Menezes, "Garment dyeing – Part – II", Colourage, January 1994, pp.29-32.

6. S. Y. Kamat & E.W.Menezes, "Machines for post-dyeing treatments", Colourage, February 1994, p.27.

7. S. Y. Kamat & E.W.Menezes, "Pretreatment", Colourage, March 1994, pp.47-49.

8. S. Y. Kamat & E.W.Menezes, "Special chemicals for pretreatment and dyeing", Colourage, April 1994, pp.59-61.

9. S. Y. Kamat & E.W.Menezes, "Dyeing of polyamide articles - I", Colourage, May 1994, pp.45-47.

10. S. Y. Kamat & E.W.Menezes, "Dyeing of polyamide articles - II", Colourage, June 1994, pp.35-37.

11. P. W. Harrison, "Garment-dyeing: Ready-to-wear Fashion from the Dyehouse", Textile Progress, Vol.19, No.2, The Textile Institute, Oxford, 1988

12. R. M. Mittal, "New Wave in Garment Exports: Garment Processing", ATIRA, Ahmedabad, 1990

An overview of functional finishing treatments on apparels and textiles

Dr. M. R. Srikrishnan[1], Dr. M. Parthiban[2]

[1]*Assistant Professor (Senior Grade), Department of Fashion Technology, PSG College of Technology, Peelamedu, Coimbatore-641004*

[2]*Assistant Professor (SG), Department of Fashion Technology, PSG College of Technology, Peelamedu, Coimbatore - 641004*

Abstract : Textiles with desired functional and aesthetic properties have an evergreen demand among the consumers globally. Unless any product is characterized by value addition, it is now impossible to survive in this highly competitive world market. Only innovative products will be sustainable to open up new markets and new horizons for the textile industry. Manufacturers should now produce products to satisfy customers that are best in terms of quality and price. Customers today have a wide range of choices and the one who produces the best quality at a better competitive price will survive and prosper. Processing is important to make a usable but finishing gives better characteristics and value addition to it. It makes textile material attractive, comfortable and finishing can incorporate desirable properties.

Keywords : Nano technology, microencapsulation, finishing, peach skin, stain resistant.

12.1   Introduction

The field of apparel and textile finishing is very broad. Globalization has added competition at the highest level. Making an apparel product more sustainable, fashionable and customer focused by increasing its both aesthetics and functional properties are the way to make the apparel products more demandable in the market. Hence, finishing is the heart of textile processing and it gives protection from soil re-deposition during laundering and absorbency or transport of liquid water.

Apparel with desired functional and aesthetic properties has an evergreen demand among the consumers globally. . Soil release finish is one of the important finishing processes applied to apparel and fabric. To make the apparel products more fashionable, sustainable, and customer focused soil release finishing processes have become a popular value addition process, which improves the different functional properties of apparel.

12.2   Value added finishes for garments

12.2.1   Ozone fading in denim finishing

With finishes, a hot concept, garment manufacturers and finishing units are innovating and experimenting with new techniques to get catchy finishes. Ozone fading in denim provides an interesting look at the garment. However, the procedure and methods to be adopted need to be carefully understood. One of the key noxious by-products of urban photochemistry is ozone and this can reach dangerously high levels of 0.5 ppm. In the presence of UV light, there is an interaction between the hydrocarbons, oxides of nitrogen and oxygen that causes the release of ozone (in addition to other compounds). The release is during the daytime due to the presence of sunlight. Indigo dyestuff tends to fade or turn yellow due to the ozone reaction. Ozone is present in the atmosphere in most industrially active and urban dwellings and is formed in the presence of sunlight or UV light. The levels of ozone can reach dangerous levels of O.5 ppm and the deteriorating effect it has on denim apparel (particularly that stored in retail shelves) is practically irreversible.

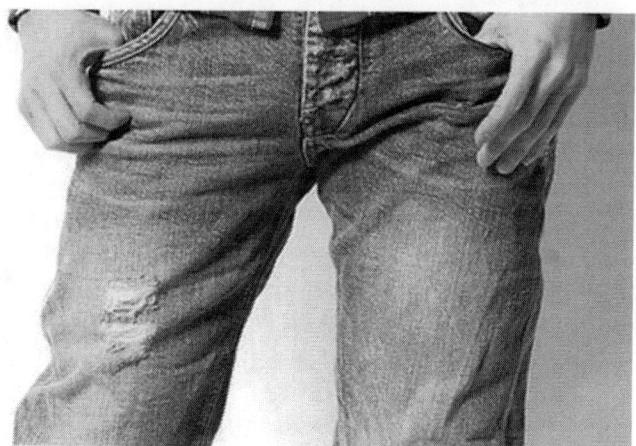

Figure 12.1: Ozone fading of denim: Waterless denim

Ozone can react with aliphatic & aromatic bonds. For this reason, it is possible to use ozone gas for an anti-felting finish of wool, without treatment becoming economically significant. Polyester is easily damaged by ozone than polyamide because polyester's aromatic ring system can be easily oxidized & destroyed by the ozone. This could be termed as ozone fading. The appearance of oxidative bleaching of blue, red, & yellow disperse dyes caused by the effect of the atmospheric zone which is similar in appearance to gas fading. This effect is most common in acetate, triacetate & polyester dyes. Heat treatment of triacetate & polyester improves fastness to ozone fading. It can be prevented by using anti-(oxidation) ozonate softening agent such as diphenylenediamine (also uses as gas fading inhibitor) & p-octylphenol (which has no gas fading inhibitor), which means that using the method detailed here, ozone fading can be prevented. The slight high cost of employing this finish is more than recovered with the improvement of performance characteristics of the denim garment. Furthermore, anti-ozonate could be a good "marketing" tool too.

12.2.2   Antimicrobial and odor control finishing for textiles

Microorganisms can be found almost everywhere in the environment. Increasing public awareness about the risks of microbial infection is a growing demand for products. The most textiles currently used in hospitals are conductive to cross infection or transmission of diseases caused by microorganisms, particularly bacteria and fungi. The medical textiles, such as fabric liners for prosthetics and casts, also require odor control. The conditions are damp and inclined to bacterial growth, which causes odor. "An antimicrobial needs power, speed, and durability to deal with those odors." The increasing need for incontinence odor control is not just confined to garments, but also is required for bedspreads, linens, and upholstery. Antibacterial fibres and variable antibacterial chemicals available in the international market are mostly from the synthetic base and are not environment-friendly. Consumer preference has changed and higher demands are placed on the functional fabrics these new requirements necessitated a production process that's environment-friendly. There are many natural plant products, which show anti-bacterial properties for e.g. extracts from roots, stem, leaves, flowers, fruits, and seeds of diverse species of plants exhibit antibacterial properties. These antibacterial extracts can be used as a textile finishing agents in the crude form or as microcapsules to enhance the durability and controlled release of the extracts.

Microbes are minute organisms, which can be most dangerous for creating harm to our lifestyle in different ways. Therefore, to make the environment healthy, hygienic, and fresh, it becomes very important to have the control overgrowth of the microbes and for these the garments/fabrics should be treated with some specialty chemicals, which can restrict the growth of these microorganisms. Antimicrobial finishing is one of the special types of finishing given to the textiles where the chances of bacterial growth are high and the safety is paramount.

Recipe: Fabshield AEM 5700 0.5 % (For exhaust method)

Fabshield AEM 5700 7.0 gpl (For padding method)

Fabshield AEM 5700 2.0 % solution (For spray method)

12.2.3   Micro Encapsulation finishing

Microencapsulation is one of the latest technologies used to impart an array of unique characteristics to a garment. Particles filled with active ingredients are applied to the fabric or garments for long-lasting effects. Microencapsulated particles are anchored onto the fiber. As the wearer moves, the capsules are activated producing a slow release of the active ingredient. Active ingredients run the gamut including moisturizers, aloe, vitamin E, therapeutic smells, and insect repellent.

Microencapsulation technology is also being used to create garments with built-in temperature control systems. Phase change materials or PCM's were originally developed for NASA to protect astronauts against temperature fluctuations ranging from bitter cold to scorching heat. Now, a regular person can enjoy the same protection. The concept is based on the endothermic/ exothermic transitions. When ice melts, heat is absorbed from the environment endothermic transition). When liquids solidify, heat is given off (exothermic transition). Phase change materials are capable of storing and releasing large amounts of energy. Microencapsulated PCM's can be applied to the fabric or garment. The PCM stores the body's excess heat as it is created, and releases it, as it is needed.

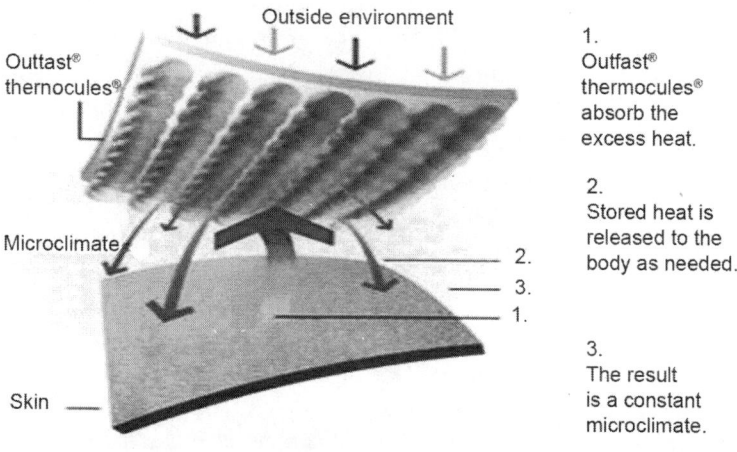

Figure 12.2: Micro encapsulation Technique

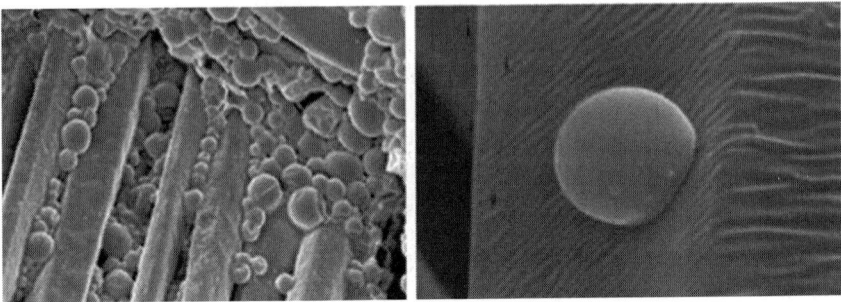

Figure 12.3: Micro encapsulation finishes

12.2.4   Stain resistant finishing

New technological advances have made the heavy stain-resistant coatings of the 1960's a thing of the past. Traditionally, fluorochemicals have been used to impart stain resistant characteristics to the textile. Fluorochemicals are the only chemicals capable of repelling water, oil and other liquids that cause stains.

Fabrics finished with fluorochemicals have nonstick properties. Unfortunately, fluorochemicals can have adverse effects on the environment and on human and animal health. This has led to the investigation of new stain-resistant finishes.

Using nanotechnology, unique and permanent stain resistant finishes are being developed. Nanotechnology is defined as the precise manipulation of individual atoms and molecules to create layered structures. Nanosize particles can exhibit unexpected properties— different from those of the bulk material. The basic premise is that properties can dramatically change when a substance's size is reduced to the nanometer range. For example, ceramics, which are normally brittle, can be deformable when their size is reduced. In bulk form, gold is inert; however, once broken down into small clusters of atoms it becomes highly reactive. Coolest Comfort is now being applied to resin-treated cotton. Resins are used to make cotton wrinkle free, unfortunately, the resin treatment also blocks cotton's natural ability to absorb moisture. Coolest Comfort can be formulated to restore the natural wicking properties of resin-treated cotton. Resists Static is the first permanent anti-static treatment for synthetic fibers. Not only does it repels static but also repels statically attractive substances such as dog hair, lint, and dust. Resists Static can be applied to a variety of fabric constructions including fleece and suit linings.

12.2.5   Nano finishing

The concept of nanotechnology was started over forty years ago and it has real commercial potential it the textile industry. The use of nanotechnology in the textile industry has increased rapidly due to its unique, valuable properties. The present status of nanotechnology use in industry is reviewed, with an emphasis on improving the properties of textile materials. The unique and new properties of nanomaterials have attracted not only the scientists and researchers but also the businesses, due to huge economic potential.

With the advent of nanoscience and technology, a new area has developed in the area of textile finishing called "Nano finishing." The impact of nanotechnology in the textile garment finishing area has brought up innovative finishes as well as new application techniques. This advanced garment finishes setup an unprecedented level of textile performances of strain-resistant, hydrophilic, antistatic, and wrinkle resistant, and shrink proof abilities and protection methods. Coating the surface of textiles and clothing with nanoparticles is an approach to the production of highly active surfaces to have UV blocking, antimicrobial, flame retardant, water repellant, and self-cleaning properties.

12.2.5.1   Application of nanotechnology in textiles

This technology is used in textile fabric finishing which helps in transforming the fabric into an entirely new carefree fabric rendering it shrinks proof, wrinkles resistant, stain and water repellent. This is intended for use in linen and cotton. Nanotechnology is used for next generation, dimension-stabilizing, and ease-of-care finish. All of these applications of nanotechnology in textile finishing are a step ahead and the methods include high-performance sky wax, waterproof sky jackets thereby allowing garments to be stain repellent, shrink proof, and wrinkle resistant.

Besides, it also helps the textile to be water repellent, to have high absorption property, protection from UV, abrasion safety, fire retardancy, colorfastness, anti-microbial functional, functional hygiene, self-cleaning and functional protection. There are nanowhiskers which makes the fabric water resistant as also makes it breathable compared to resin finishes. This technology also provides oil and water repellency along with superior durability which maintains the fabric in a soft and natural wrinkle resistance state. The nanonet injects linen property and covers the core fibers in the synthetic fibers completely which helps in the linen absorption capacity of the polyester fibre. It can render a feel of linen and cotton that gives a cooling effect by absorbing the body moisture through the alteration of synthetic fibres.

12.2.5.2   Nanotechnology in textile finishing

What is sol-gel processing?

It is a process for making very small particles 20 to 40 nm that is virtually impossible to make by conventional grinding. Its main use at present seems to be for optical coatings where the finer particles give better optical clarity. Manufacture of fine a ceramic fibre seems to be the other common application.

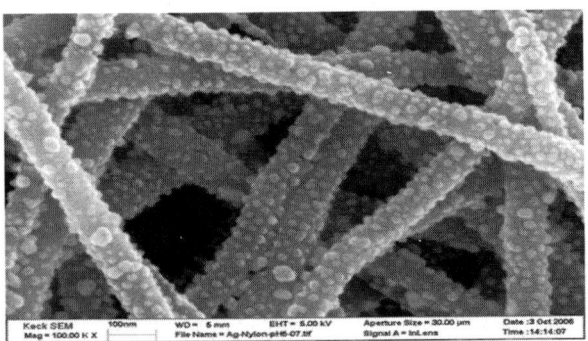

Figure 12.4: Nano particles embedded in fibre structure

How does sol-gel processing work?

A liquid precursor of the particle is dissolved in a solvent, usually, alcohol, water is added and then acid or base. The mixture is coated or cast. The precursor then decomposes to form the fine ceramic particles. If the particle concentration is high enough, the mixture gels. The gel is dried and then heated at high temperature to sinter the ceramic, giving the desired ceramic film or fibre. During this drying and sintering process, shrinkage occurs through loss of solvent and air, and this shrinkage must be carefully controlled to avoid cracking.

UV protective finish

The most important functions performed by the garment are to protect the wearer from the weather. However, it is also to protect the wearer from harmful rays of the sun. The rays in the wavelength region of 150 to 400 nm are known as ultraviolet radiations. The UV-blocking property of a fabric is enhanced when a dye, pigment, delustrant, or ultraviolet absorber finish is present that absorbs ultraviolet radiation and blocks its transmission through a fabric to the skin.

Metal oxides like ZnO as UV-blocker are more stable when compared to organic UV-blocking agents. Hence, nano ZnO will really enhance the UV-blocking property due to their increased surface area and intense absorption in the UV region. For antibacterial finishing, ZnO nanoparticles scores over nano-silver in cost-effectiveness, whiteness, and UV-blocking property. Fabric treated with UV absorbers ensures that the clothes deflect the Harmful ultraviolet rays of the sun, reducing a persons UVR exposure and protecting the skin from potential damage. The extent of skin protection required by different types of human skin depends on UV radiation intensity & distribution in reference to geographical location, time of day, and season. This protection is expressed as SPF (Sun Protection Factor), higher the SPF Value better is the protection against UV radiation.

12.2.6   Peach skin effect

In classical finishing of lyocell, peach Skin effects are produced in rope form using Airflows and Airtumblers. Alternatives are produced in open- width finishing with special emery finishes. Here for example, emery papers coated with diamond dust are recommended which imitate effects comparable with the classical method. The peach skin effects can also be achieved with the treatment of cellulose enzymes where one does the fibrillation with alkali and later treatment with the cellulose enzymes that polish the fibrils to produce a peach skin permanent effect. Apart from Peach Skin effects, various finishes can naturally be applied to fabrics of Lenzing Lyocell such as calendaring, embossed effects, crepe or other special effects.

12.3   Conclusion

Globalization has opened the doors to competition at the highest level. Every industry must now produce products that are best in terms of quality & price. Customers now have a wide variety of choices in each price range & one who offers the best at a competitive price will survive & prosper. For the Indian Textile Industry, challenges are even greater. Technologically speaking in the textile processing industry there is still a lot to ask for. The thrust areas in textiles, in the coming years, will very much center on process optimization through improved shorter duration processes i.e. efficient & effective processes, customized solutions, value addition, global competitiveness: high standards of quality at lowest possible cost, and eco awareness.

New innovative functional finishes represent the next generation of finishing industry, which, make textile materials act by themselves. As with all emerging technologies, a successful future for Functional Finished Textiles will only be achieved through open sharing of ideas and research findings, a thorough testing of the capability boundaries, and frank discussion of fears and failings. India has a bright future in Functional Finished Textiles.

12.4   References

1. K.V.Lokesh, "Value addition finishes for textiles", Textile learner, 2018

2. Mark H., Wooding N.S. and Wiley S.M., Chemical After treatment of Textiles; Wiley: New York, 1971 Retrieved, 520 (2006)

3. Md. Mazedul Islam and Adnan Maroof Khan, "Functional Properties Improvement and Value Addition to Apparel by Soil Release Finishes - A General Overview", Research Journal of Engineering Sciences, Vol. 2(6), 35-39, June (2013), ISSN 2278 – 9472.

4. M.Parthiban, S Ramanathan, S Nithyananthan, J Balakrishnan & B Manimaran, "A Comprehensive Overview of Antimicrobial & Odor Control Finishing For Textiles" "Technical Textiles",

5. Yuke O., Test Method of Antimicrobial Finished Fabrics (in Japanese), "Antimicrofinish" The Antimicrobial Finish Society of Japan Textiles Co., 1989 pp 182-184.

6. K. K. Sonue & M. G. Labhe, 1982, Colourage, vol. 29 p27

7. Buschle-Diller & Yang ,2001,Textile Research Journal, p 388

8. D. P. Chakraborty & J. K. Sharma , 1999, Indian J. Fibre & Textile Res. Vol.24, p 120

9. R. Venkatraj , 1987, Ph.D thesis , IIT , New Delhi, p 40

10. R. M. Mittal, M. L. Gulrajni, 1988, Amer. Dyes. Rep.vol. 77, p 20

11. M. Prabharan, R. C. Nayar & J. V. Rao, 2000, Text. Rresearch J. Vol .21, p 19

12. J. W. Rucker, 1989, Text. Chemist & Colorist, vol. 70, p 657

13. Edward Menezes, nov. 2003, National Textile Conference, p 14-15.

14. https://clothingindustry.blogspot.com/2017/12/ozone-fading-denim.html

15. Textile School Last updated Mar 12, 2018

16. "Priyadarshinee Nath", "Nano finishing in textiles – an Introduction", OCS, - Online clothing study, 2019.

17. Functional textiles - www.empa.ch

18. Xin J H, Daoud W A and Kong Y Y: A New Approach to UV-Blocking Treatment for Cotton Fabrics, Textile Research Journal, 2004. 74: pp 97-100.

19. V. Parthasarathi, "Nano technology adds value to textile finishing", Indian Textile journal, Jan 2008.

20. Kathiervelu S S: Applications of Nanotechnology in Fibre Finishing, Synthetic Fibres, 2003. 32: pp 20-22.

21. P. Ganesan and L. Sasikala, " Functional finishes for Apparels", fibre 2 fashion.com

22. Mark, H.; Wooding, N. S.; Wiley, S. M. Chemical After treatment of Textiles; Wiley: New York, 1971; p 520.

23. "Chemical processing of fibers and fabrics - functional finishes", Part B, edited by M. Lewin and S. B. Sello, Marcel Dekker, New York, 1984, 515 pp.

13

Softening of woven jute cotton mixture fabrics

Dr. R. Prathiba Devi,

Assistant Professor (SG), Department of Apparel and Fashion Design,
PSG College of Technology, Coimbatore-641004, Email: prathiba17@gmail.com

Abstract : The Jute/Cotton mixture fabrics are taken up for the study. The fabrics are treated with enzymes followed by silicone polyurethane softening finish. The treated fabrics are subjected to their SEM and FTIR analysis. The results showed that there is a significant improvement in the surface softness of the jute fiber. The FTIR results confirmed the presence of softeners on the fibre surface. The experiment results proved that there is a significant improvement in wicking, water absorbency, air, water vapour permeability, and thermal conductivity. Silicon polyurethane finish brings in significant changes in the jute/cotton fabrics and the fabric improves the wear life of the garment without affecting the comfort properties.

Keywords : Comfort properties, FTIR, Enzyme treatment, Jute/cotton mixture fabric, Silicon-polyurethane

13.1   Materials and Methods

13.1.1   Materials

The textile materials and chemicals used for this research work and the experimental design of the research is given below.

13.1.1.1   Textile fibre substrate

Jute cotton mixture fabrics with cotton yarn in the warp and jute cotton mixture yarns of different ratio in the weft were used in the study. The specifications of the jute cotton mixture fabrics used in this study are as follows:

(a)   Jute cotton mixture fabric (30/70): warp count – 30^s Ne, weft count - 8^s Ne, ends per inch – 44, picks per inch – 44, GSM – 570

(b)   Jute cotton mixture fabric (40/60): warp count – 30^s Ne, weft count - 4^s Ne, ends per inch – 48, picks per inch – 42, GSM – 623

(c)   Jute cotton mixture fabric (50/50): warp count – 30^s Ne, weft count - 6^s Ne, ends per inch – 48, picks per inch – 42, GSM – 643

(d)   Jute cotton mixture fabric (70/30): warp count – 30^s Ne, weft count - 8^s Ne, ends per inch – 38, picks per inch – 38, GSM – 710

13.1.1.2   Chemicals and auxiliaries

The chemicals such as sodium chloride, sodium carbonate, sodium hydroxide, sodium hypochlorite, hydrogen peroxide, hydrochloric acid, acetic acid, hot brand reactive dye, cellulase enzyme, polymethylsiloxane silicone softener, titanium dioxide, and chitosan mentioned elsewhere in this study were of analytical (AR) grade.

13.1.2   Methods

13.1.2.1   Pretreatment of jute cotton mixture fabrics

The woven textile jute cotton mixture fabrics were given pretreatment process like desizing, scouring, bleaching and mercerisation as mentioned by Karmakar.

13.1.2.2   Enzymatic treatment of jute cotton mixture fabrics

After the pre-treatment, the jute cotton mixture fabrics were treated with an enzyme to impart softness. Suggest a mixed enzyme product comprising cellulose, xylanase, and pectinase, which can react with cellulose, hemicellulose and pectin of lignocellulosic material to soften it, as this combination is eco-friendly. Hence, cellulase enzyme was selected and used as mentioned in the recipe given below:

Cellulase enzyme – 5% (owm)

Acetic acid – 0.5% (owm)

Temperature - 55°C

Time – 30 minutes

MLR – 1:20 (owm – on weight of material)

13.1.2.3   Dyeing of jute cotton mixture fabrics

The enzyme-treated jute cotton union fabrics were dyed with hot brand reactive dyes by exhaust method. The dye (3% owf) and sodium chloride (30g/l) were pasted with water at 45 ° C and dissolved by adding water at 80° C and kept for 30 min with material liquor ratio at 1:20. Then sodium carbonate (1% owf) was added for fixation of dyes and run for 1 hour at 60° C. The pH maintained was 5-7. Finally, the fabrics were hot washed, soaped, washed, and neutralized with acetic acid.

13.1.2.4   Silicone-polyurethane finishing of jute cotton mixture fabrics

The amino-functional based polymethylsiloxanes silicone softener treatment was given by Pad-dry-cure method on the reactive dyed jute cotton mixture fabrics, with 10 gpl of amino silicone softener and 10 gpl of Polyurethane solution at pH 6.0 (maintained by acetic acid) and temperature 40°C for 15 minutes with the pressure of 1 kg/cm^2 in order to get optimum pick-up of 0.8% owm. Then the fabrics were dried at 100°C for 3 min and cured at 150°C for 4 min in the drying and curing chamber respectively.

13.2   Studies on characterisation and evaluation of silicone polyurethane finished jute cotton mixture fabrics

This chapter deals with the physical characterisation and evaluation of enzyme and silicone finished jute cotton mixture fabrics.

13.2.1   Tensile strength of jute cotton mixture fabrics

The tensile properties of jute cotton mixture fabrics of different proportions are provided in table 13.1 and figure 13.1. From the table is clear that the tensile strength of the unfinished warp yarns appears to be higher as the jute proportion increases. As the jute fiber has good strength, the tensile strength of the weft yarns also increased, as there is an increase in jute proportion. However, after finishing treatment there is a reduction in the tensile strength of the warp and the weft yarns. This may be attributed to the length of the fibers, optimum GSM of the fabrics and various chemical pretreatments.

Table 13.1: Tensile Strength of the finished and unfinished jute cotton mixture fabrics

S.No.	Name of the sample	Tensile strength (kg)			
		Unfinished fabric		Finished fabric	
		Warp	Weft	Warp	Weft
1	Jute/Cotton 30/70	85	95	63	45
2	Jute/Cotton 40/60	133.85	65	96.25	50
3	Jute/Cotton 50/50	115.25	122.5	85.25	98.5
4.	Jute/Cotton 70/30	100	110	74	64

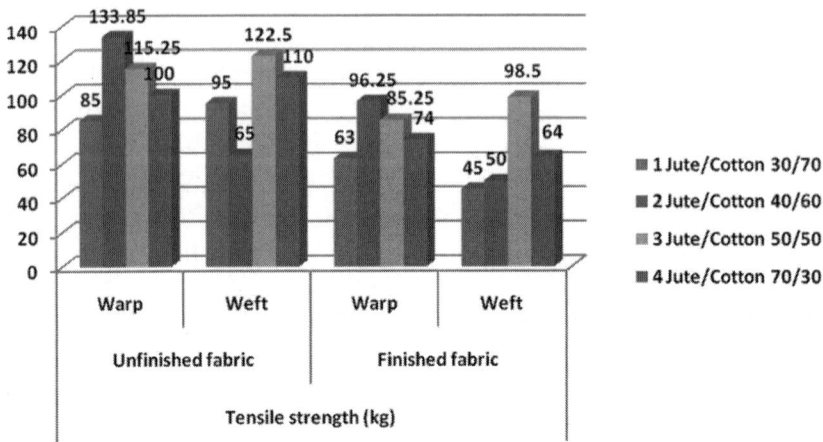

Figure 13.1: Tensile Strength of the finished and unfinished jute cotton mixture fabrics

13.2.2   Tearing strength of jute cotton mixture fabrics

Table 13.2 and Figure 13.2 reveal the tearing strength of the unfinished and finished jute cotton mixture fabrics. The warp way and weft way tear strengths of the finished fabric samples decrease after finishing. From the table, it is depicted that the jute cotton mixture fabric was higher and better than 30/70 and 40/60 jute cotton mixture proportions which are contributed by high tensile strength property of the jute fibers and high GSM.

Table 13.2: Tearing Strength of the finished and unfinished jute cotton mixture fabrics

S. No.	Name of the sample	Tearing strength (kg)			
		Unfinished fabric		Finished fabric	
		Warp	Weft	Warp	Weft
1	Jute/Cotton 30/70	2.500	3.904	2.411	3.302
2	Jute/Cotton 40/60	5.783	1.99	4.88	1.912
3	Jute/Cotton 50/50	6.543	5.1	4.16	3.808
4.	Jute/Cotton 70/30	3.306	4.672	2.990	3.475

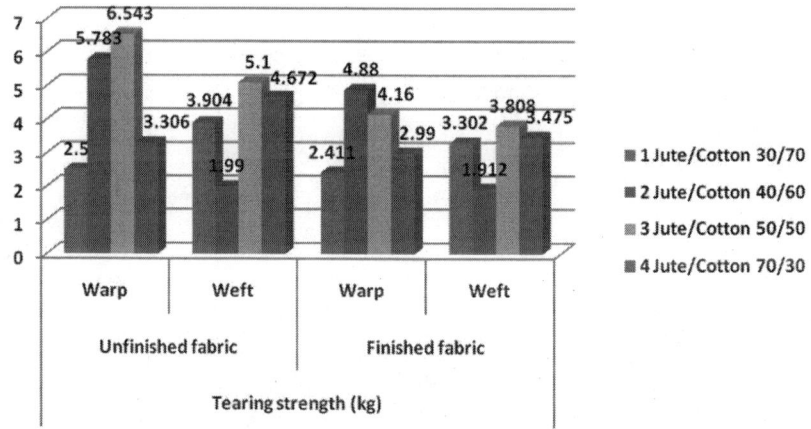

Figure 13.2: Tearing Strength of the finished and unfinished jute cotton mixture fabrics

13.2.3   Stiffness test of jute cotton mixture fabrics

13.2.3.1   Flexural Rigidity of jute cotton mixture fabrics

Table 13.3 and Figure 13.3 show the flexural rigidity of the fabric samples before and after finishing. Table 3 reveals a decrease in flexural rigidity or stiffness of the jute cotton mixture samples after finishing which indicates that the samples became more flexible after finishing treatment. This flexibility developed due to swelling and loss of lignin, hemicelluloses of jute. This swelling property was improved by breaking hydrogen bonds in the fiber cellulose molecule. This develops the soft handling properties of the jute materials.

Table 13.3: Flexural Rigidity of the finished and unfinished jute cotton mixture fabrics

S. No.	Name of the sample	Flexural Rigidity (mgcm)			
		Unfinished fabric		Finished fabric	
		Warp	Weft	Warp	Weft
1	Jute/Cotton 30/70	2147.40	9941.6	1019.29	4083.5
2	Jute/Cotton 40/60	387.69	958.14	198.30	716.38
3	Jute/Cotton 50/50	315.20	713.2	268.78	658.67
4.	Jute/Cotton 70/30	1419.0	12383.5	1319.0	5284.6

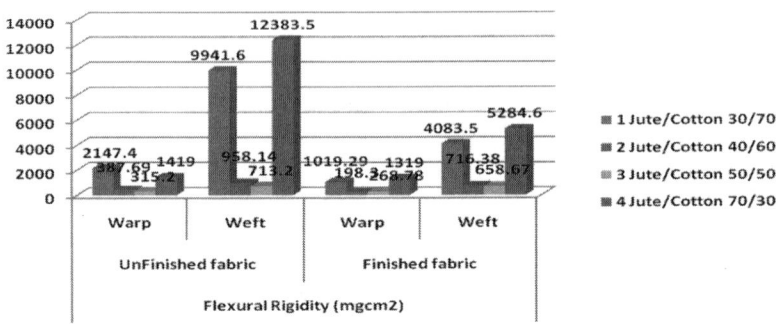

Figure 13.3: Flexural Rigidity of the finished and unfinished jute cotton mixture fabrics

13.2.3.2   Bending Modulus of jute cotton mixture fabrics

Table 4 and Figure 13.4 show the bending modulus of the jute cotton mixture fabric samples. Bending modulus determines the ability of the fabric to bend under its own weight to a definite extent. From table 13.4 it is clear that the unfinished jute cotton fabrics of 50/50 and 70/30 shows greater bending modulus due to greater jute content, which is a noted disadvantage of jute fibers. This rigidity was reduced to some extent by treating the fabrics with enzymes during pretreatment and silicone polyurethane in the final stage.

The finished samples of 50/50 and 70/30 proportions showed comparatively lesser bending modulus after finishing. Overall bending modulus of all the samples was decreased after finishing. From table 13.4 it is clear that flexural rigidity of the unfinished fabric is more, which leads to higher bending modulus of the unfinished samples. In addition, the flexural rigidity or stiffness of the finished samples reduced after finishing treatment, which resulted in lesser bending modulus. Higher the stiffness of the fabric higher is the bending length. Bending modulus is a measure of stiffness that determines the draping quality.

Table 13.4: Bending Modulus of the finished and unfinished jute cotton mixture fabrics

S. No.	Name of the sample	Bending Modulus (kg/cm²)			
		Unfinished fabric		Finished fabric	
		Warp	Weft	Warp	Weft
1	Jute/Cotton 30/70	26.47	122.6	16.5	66.14
2	Jute/Cotton 40/60	76.75	23.91	45.05	14.37
3	Jute/Cotton 50/50	119	98.96	47.28	87.75
4.	Jute/Cotton 70/30	17.36	134.36	15.32	69.59

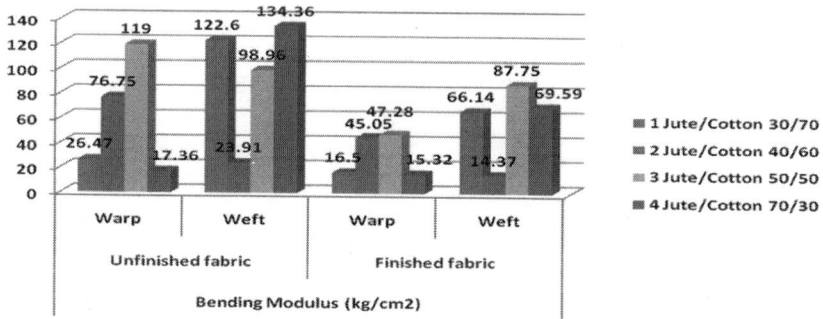

Figure 13.4: Bending Modulus of the finished and unfinished jute cotton mixture fabrics

13.2.4   Drapeability of jute cotton mixture fabrics

The drapeability of the fabric samples are found in table 13.5 and figure 13.5. Table 13.5 shows that the drape coefficient of the unfinished jute cotton mixture fabrics increased gradually from 0.72% for 30/70, 0.82% for 40/60, 0.87% for 50/50 and 0.89% for 70/30 as the jute proportion in the fabrics increases. Higher the drape coefficient stiffer is the fabric which is also attributed by the great thickness leading to high stiffness characteristics of the jute fibers. This stiffness leads to less drapeability. However from the figure 5, it is clear that after silicone polyurethane finishing treatment the softness of the jute cotton mixture fabric led to a reduction in drape coefficient in all the proportions significantly, leading to increase in drapeability of all the jute cotton mixture fabric proportions.

Table 13.5: Drapeability of the finished and unfinished jute cotton mixture fabrics

S. No.	Name of the sample	Drapeability (%)	
		Unfinished fabric	Finished fabric
1	Jute/Cotton 30/70	0.72	0.25
2	Jute/Cotton 40/60	0.82	0.63
3	Jute/Cotton 50/50	0.87	0.54
4.	Jute/Cotton 70/30	0.89	0.42

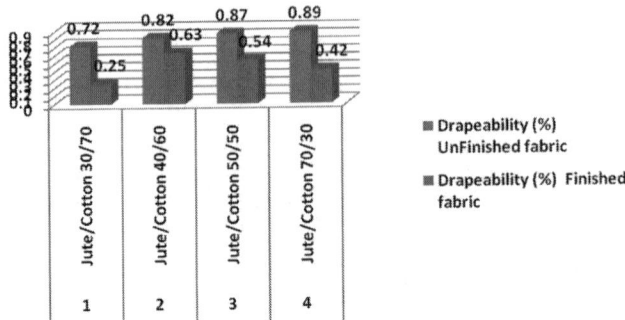

Figure 13.5: Drapeability of the finished and unfinished jute cotton mixture fabrics

13.2.5   Pilling of jute cotton mixture fabrics

Table 13.6 and Figure 13.6 show the resistance of test samples for pilling. Among the fabric samples before and after finishing the unfinished samples showed moderate surface fuzzing with rating 4. Excellent resistance for pilling was observed in the samples treated with enzyme and silicone polyurethane finishes with rating 5 and no visual change.

Table 13.6: Pilling of the finished and unfinished jute cotton mixture fabrics

| S. No | Name of the Sample | Pilling | |
		Unfinished fabric	Finished fabric
1	Jute/Cotton 30/70	4	5
2	Jute/Cotton 40/60	4	5
3	Jute/Cotton 50/50	4	5
4.	Jute/Cotton 70/30	4	5

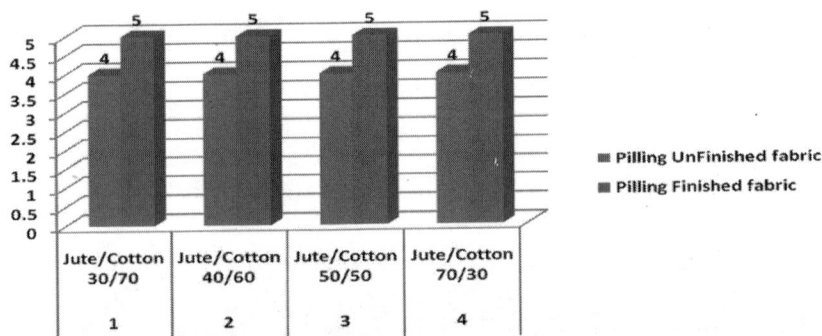

Figure 13.6: Pilling of the finished and unfinished jute cotton mixture fabrics

13.2.6   Wickability of jute cotton mixture fabrics

Table 13.7 and figure 13.7 shows the wickability height of both finished and unfinished jute cotton mixture fabrics. When compared to the Unfinished fabric wickability of Finished fabric is improved significantly ($P>0.05$) 20%, 30%, 31%, and 40% respectively for 30/70, 40/60, 50/50 and 70/30 Jute/ Cotton mixture. Among these samples, the fabric with more jute proportion showed higher wickability. The Optimum wickability percentage was found in 40/60 and 50/50 jute cotton mixture proportions. This is because of the application of hydrophilic softener which increases the capillary flow in the fabric. The removal of hydrophobic noncellulosic components from the fabric surface by enzyme treatment is another reason for the improved wicking.

Table 13.7: Wickability of the finished and unfinished jute cotton mixture fabrics

S. No	Name of the Sample	Wickability (cm)	
		Unfinished fabric	Finished fabric
1	Jute/Cotton 30/70	5.01	6.01
2	Jute/Cotton 40/60	5.7	7.4
3	Jute/Cotton 50/50	4.9	6.4
4	Jute/Cotton 70/30	5.15	7.2

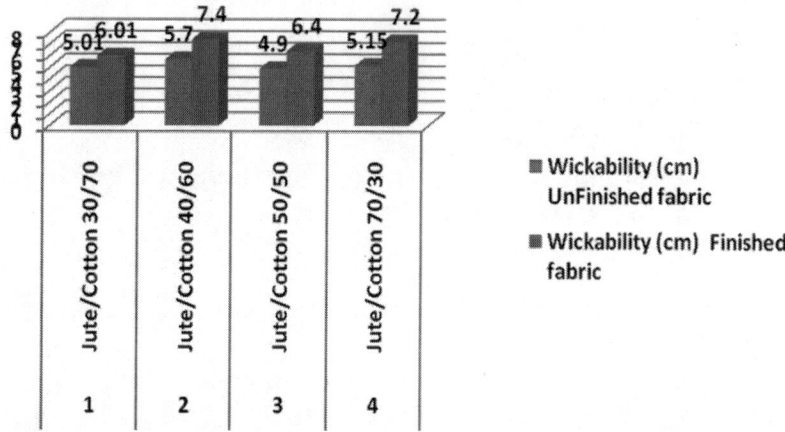

Figure 13.7: Wickability of the finished and unfinished jute cotton mixture fabrics

13.2.7   Water absorbency of jute cotton mixture fabrics

From the table 13.8 and figure 13.8 it is clear that the water absorbency of Finished fabric has improved (P>0.05) 10% for 30/70, 12% for 40/60, 17% for 50/50 and 31% for 70/30 Jute /Cotton mixture proportions. As the jute proportion increases, the absorbency also increases. This is due to the application of silicone and polyurethane treatment, which improves the swelling of the fiber, which increases the accessibility area leading the resulting fiber to have higher water absorbency.

This will give more comfort to the wearer when the garment is worn next to the skin. The cellulase enzyme is able to gain access to the cellulose, and in due process, removes hydrophobic noncellulosic components from the fabric surface thus the treatment of enzyme also helps to improve the water absorbency of finished material.

Table 13.8: Water Absorbency of the finished and unfinished jute cotton mixture fabrics

S. No	Name of the Sample	Water Absorbency (sec)	
		Unfinished fabric	Finished fabric
1	Jute/Cotton 30/70	4 sec	3.6 sec
2	Jute/Cotton 40/60	5 sec	4.8 sec
3	Jute/Cotton 50/50	6 sec	5 sec
4	Jute/Cotton 70/30	6.5 sec	4.5 sec

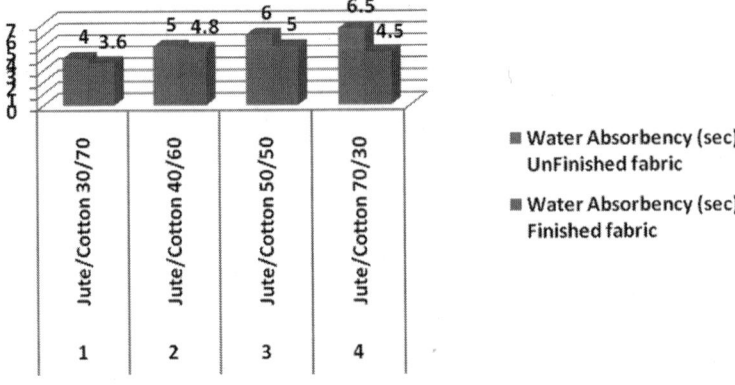

Figure 13.8: Water Absorbency of the finished and unfinished jute cotton mixture fabrics

13.2.8   Water vapour permeability of jute cotton mixture fabrics

Table 13.9 and figure 13.9 shows the Water Vapour Permeability of both unfinished and finished Jute/Cotton union fabrics. Water vapor permeability is an important parameter in evaluating comfort characteristics of the fabric, as it represents the ability to transfer perspiration. The Water Vapour Permeability of Finished fabric has increased significantly ($P>0.05$) up to 10%, 16%, 14%, and 13% respectively for 30/70, 40/60, 50/50 and 70/30 proportions of jute cotton mixture fabric.

Two mechanisms can be considered for water vapour transfer through the fabric: one is through fabric pores and the second is through absorption by fabric and then evaporation from fabric surface. The second mechanism is applied here, the increase in water vapour permeability because of the softening finish applied which increases the capillary transfer of water vapour through fibre bundles, and the surface morphology is modified through the finish and enzyme treatment, which reduces the roughness and hairiness of the fabric. This will increase the moisture vapour transmission through the fabric.

Table 13.9: Water Vapour Permeability of the jute cotton mixture fabrics

S. No	Name of the Sample	Water Vapour Permeability (%)	
		Unfinished fabric	Finished fabric
1	Jute/Cotton 30/70	7.8	8.6
2	Jute/Cotton 40/60	6.7	7.8
3	Jute/Cotton 50/50	5	5.7
4.	Jute/Cotton 70/30	4.6	5.2

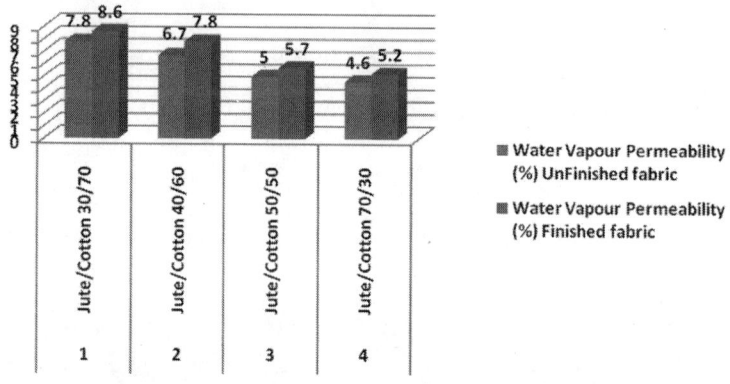

Figure 13.9: Water Vapour Permeability of the jute cotton mixture fabrics

13.2.9   Air permeability of jute cotton mixture fabrics

Table 13.10 and figure 13.10 shows the Air Permeability of both finished and unfinished jute cotton mixture fabrics. The Air Permeability of Finished fabric has significantly increased by 5.2 cm in 30/70, 40.2 cm in 40/60, 27.5 cm in 50/50 and 22.6 in 70/30 jute cotton mixture fabric proportions. This is because of the softening treatment, which modifies the fabric surface by reducing roughness, and hairiness which in turn will increase the space between the yarns. This fabric will have more air transfer due to the clean and smooth fiber edges. The application of different chemical treatments results in slight openness of the fabric, which will increase the air permeability. This will enhance the comfort of the garment.

Table 13.10: Air Permeability of the jute cotton mixture fabrics

S. No	Name of the Sample	Air Permeability ($cm^3/cm^2/s$)	
		Unfinished fabric	Finished fabric
1	Jute/Cotton 30/70	54.1	59.3
2	Jute/Cotton 40/60	38.9	79.1
3	Jute/Cotton 50/50	23.9	51.4
4	Jute/Cotton 70/30	42.9	65.5

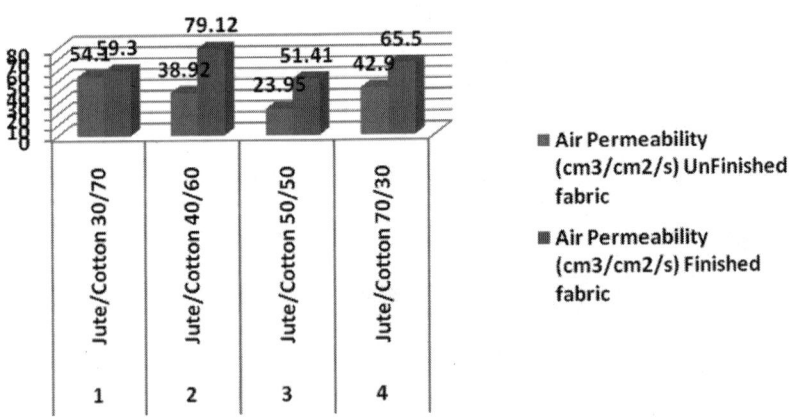

Figure 13.10: Air Permeability of the jute cotton mixture fabrics

13.2.10   Thermal conductivity of jute cotton mixture fabrics

Table 13.11 and figure 13.11 shows the Thermal Conductivity Coefficient of both finished and unfinished jute cotton mixture fabrics. The Thermal

Conductivity of finished fabric increased significantly (P>0.05) by 72% for 30/70, 30% for 40/60, 43% for 50/50 and 200% for 70/30 jute cotton mixture proportions. This increase in the conductivity is because of weight loss in the finished fabric, which in turn allows more heat to pass through the fabric. This will increase the water absorbency because whenever fibres absorb liquid water or vapour, heat will be released.

Table 13.11: Thermal Conductivity of the jute cotton mixture fabrics

S. No	Name of the Sample	Thermal Conductivity (W/mk)	
		Unfinished fabric	Finished fabric
1	Jute/Cotton 30/70	0.0057	0.0098
2	Jute/Cotton 40/60	0.0044	0.0057
3	Jute/Cotton 50/50	0.0037	0.0053
4.	Jute/Cotton 70/30	0.0026	0.0078

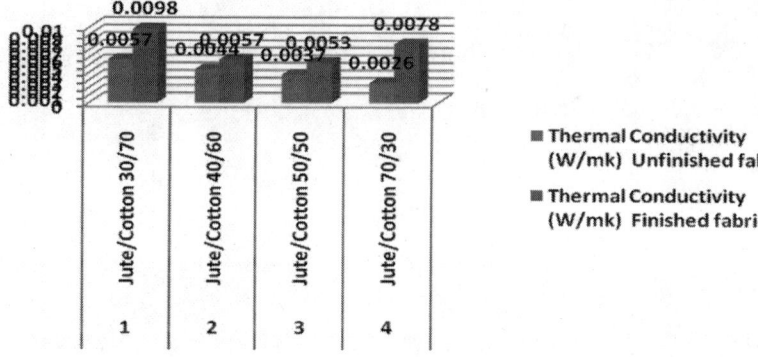

Figure 13.11: Thermal Conductivity of the jute cotton mixture fabrics

13.2.11   SEM analysis of jute cotton mixture fabrics

SEM micrographs of Finished and Unfinished fabrics taken at the magnification of 500 X are shown in the following figures. The primary function of fabric softening is to make the fabric surface soft. Figures 13.12, 13.13 & 13.14 show the fibre coating and smoothing of the rough edges of the fibres. It is also observed that the finished fabric has smooth white layers from the silicone-polyurethane finishing. The polymer film is the main cause for changes on the surface and mechanical properties of the silicone-Finished fabrics. The formation of the polymer film increases the uniformity of the surface (while MMD decreases) and decreases the roughness.

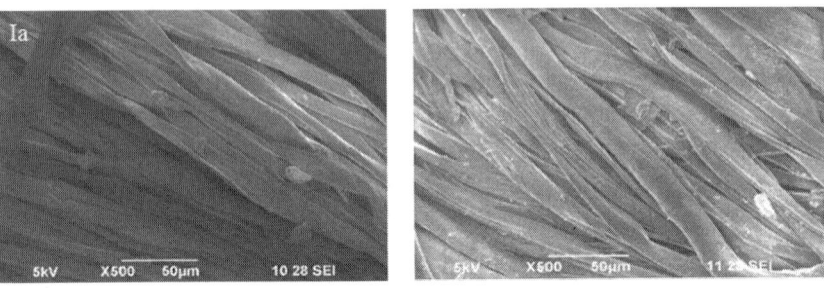

Figure 13.12: SEM micrographs of unfinished & finished 30/70 Jute Cotton mixture fabric

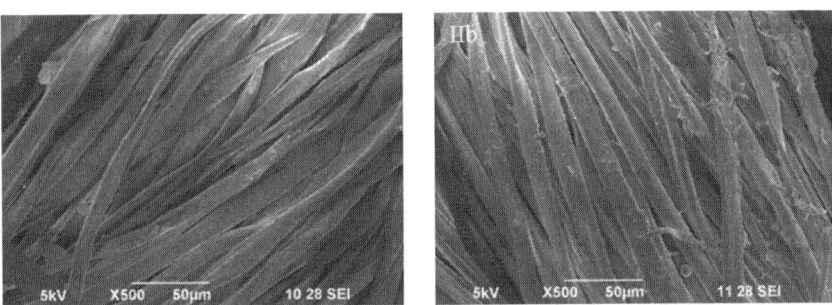

Figure 13.13: SEM micrographs of unfinished & finished 40/60 jute cotton mixture fabric

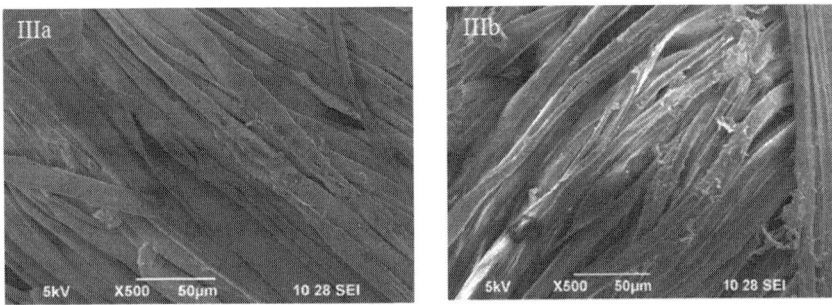

Figure 13.14: SEM micrographs of unfinished & IIIb finished 50/50 jute cotton mixture fabric

13.2.12   FTIR analysis of jute cotton mixture fabrics

Figures 13.15, 13.16, 13.17 & 13.18 shows the FTIR spectrum of the 30/70, 40/60, 50/50 and 70/30 silicone polyurethane finished jute cotton mixture fabrics respectively. From the spectra, it is confirmed that the finished fabric shows evidence of the deposition of silicone and polyurethane substance onto the surface. The absorption peak in the region 1000 cm^{-1}to 1100 cm^{-1} for the finished sample confirms the presence of -Si-O and Si-O-Si group [136,262]. This wavelength indicates the presence of silicone softeners on the fabric.

The absorption peak in the region of 3402 cm^{-1} represents the - NH stretching of the urethane group. The absorption at the region 1593 cm^{-1} represents the -N-H bonding and 1786 cm^{-1} indicate the formation of C=O stretching (free) of the urethane group. These wavelengths indicate the presence of polyurethane on the surface of the finished fabric.

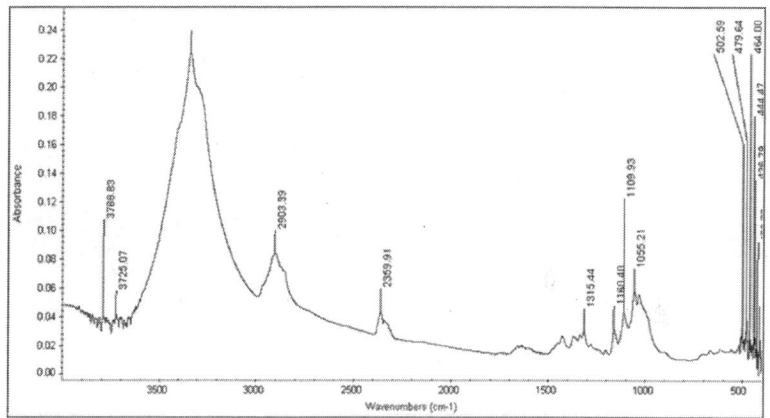

Figure 13.15: FTIR graph of finished 30/70 jute cotton mixture fabric

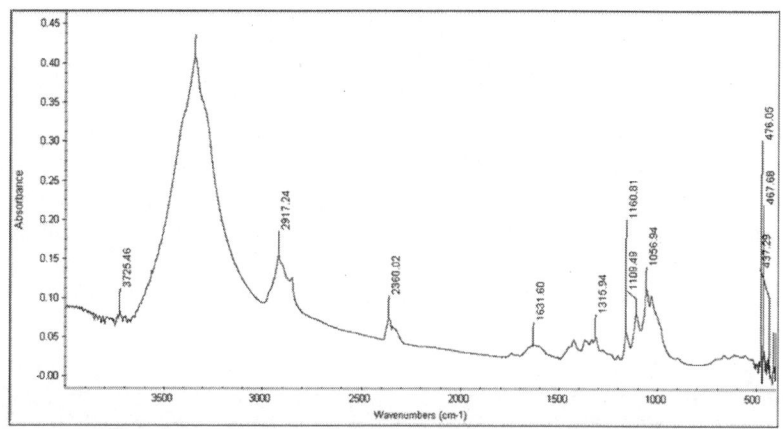

Figure 13.16: FTIR graph of finished 40/60 jute cotton mixture fabric

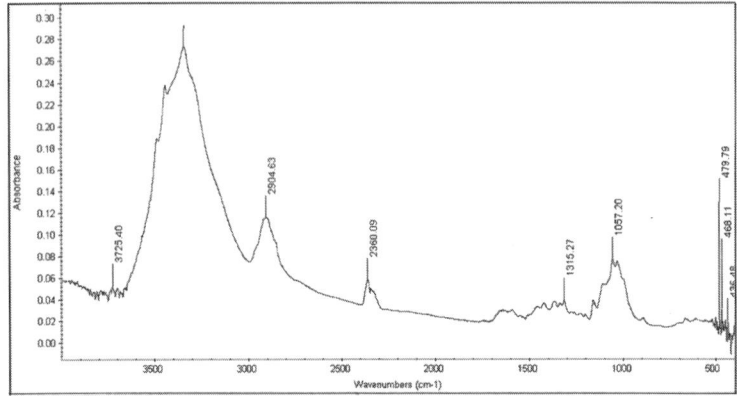

Figure 13.17: FTIR graph of finished 50/50 jute cotton mixture fabric

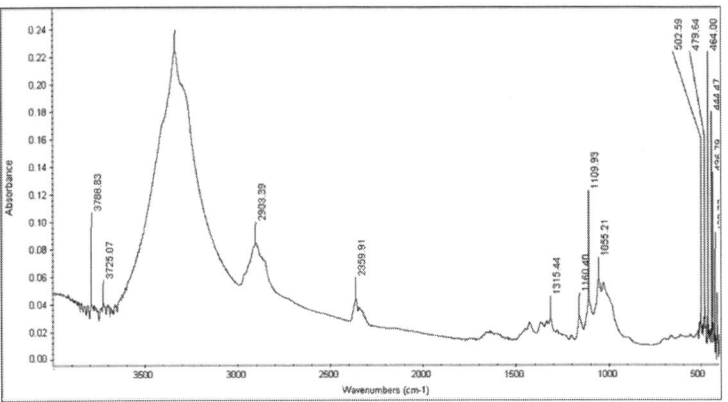

Figure 13.18: FTIR graph of finished 70/30 jute cotton mixture fabric

13.3   Conclusions

The Jute/Cotton mixture fabrics are given enzyme treatments followed by silicone polyurethane softening the finish. The SEM analysis indicates the improved surface softness of the jute fiber compared to the raw fabric. The FTIR results confirmed the presence of softeners on the fibre surface. The experiment results were evident that there is a significant ($p<0.05$) improvement in wicking, water absorbency, air, and water vapour permeability and thermal conductivity of finished fabric when compared with the raw fabric. This improvement is achieved by the enzymatic processing and finishing of textile with a silicon polyurethane finish. This shows promising results that jute can also be incorporated in our day-to-day life clothing without affecting the comfort properties. However, the scope of this research is identified as increasing the percentage of jute and evaluating their comfort properties and in the economic viewpoint.

13.4   References

1. Abul Kalam Azad, S. M., Golam, Khan, A. H., Badier Rahman, S. M., (1646) Studies on the Physical Properties of Jute-Cotton Blended Curtain and 100% Cotton Curtain, Journal of Applied Science, Voulme 7, Issue 12, p.1643

2. Chinta, S.K., Pooja, D., Gujar, (2013), Significance of Moisture Management in Textiles, International Journal of Innovative Research in Science, Engineering and Technology Vol. 2, Issue 6, Pp 2104 – 2114.

3. Ding, X. M., Hu, J. L., Tao, X. M., (2005), Free Volume and Water Vapour Transport Properties of Temperature-Sensitive Polyurethane, J. Polym. Sci. B. 43(14), Pp. 1865-1872.

4. Doke, S. S., Mandhaniya, V., & Balsubramanya, R. H., (1999), Cotton Jute Blended Textiles, International Seminar on Cotton and its Utilization in the 21st Century, CIRCOT, Indian Society for Cotton Improvement, Mumbai, Pp 56.

5. Fourt, L., and Harrist, M., (1947), Diffusion of water vapor through textiles, Textile Research Journal, Vol.17, Pp. 256-263.

6. Hsieh et al, US patent 6, 066, 494, May (2000).

7. Hunter, L., Kawabata, S., Gee, E. & Niwa. M. (1983), 'The Effect of Wool Fibre Diameter and Crimp on the Objectively Measured Handle of Woven Fabrics', 2nd Symposium on Objective Specification of Fabric Quality, Mechanical Properties and Performance, organised by the Text. Mach. Soc. of Japan, Osaka. pp. 167.

8. Ian Holme, (2007), Innovative Technologies for High Performance Textiles, Society of Dyers and Colourists, Color. Technol., 123, 59–73

9. Kang T J & Kim M S, 2001, 'Effects of Silicone treatments on the dimensional properties of the wool fabric', Textile Research Journal, vol. 71, no. 4, pp. 295-300.

10. Mythili, C. V. Malar Retna, A. & Gopalakrishnan, S. (2003), 'Synthesis, mechanical, thermal and chemical properties of polyurethanes based on cardanol', Bull. Mater. Science, vol. 27, no. 3, pp. 235-241.

11. Noll W, 1968. 'Chemistry and Technology of silicones', pp. 663-683.

12. Parthiban M and Ramesh Kumar M, Indian Journal of Fibre and Textile Research, 32, 446-452, (2007).

13. Prasad, A.K., (2007), Novel Effects in Garment Processing and Value Added Finishes, Journal of the Textile Association, Pp 39-42.

14. Shridhar V. Chikkodi, Samina Khan and R .D. Mehta, Textile Research Journal 65, 564, (1995).

15. Tzanov, T.Z., Betcheva, R., Hardalov, I & Hes, L. (1998). Quality Control of Silicone Softener Application', Textile Research Journal, vol. 68, pp. 749.

16. Walter N, 1968. —Chemistry and Technology of silicones□, p 663-683.

Combination herbal extract treated healthcare apparel for common skin diseases

K. Chandrasekaran[1*], M. Senthilkumar[2]

[1*]*Department of Fashion Technology, PSG College of Technology, Coimbatore – 641 004*
[2]*Department of Textile Technology, PSG College of Technology, Coimbatore-641 004*
Email*: kcs.fashion@psgtech.ac.in

Abstract : This work is aimed to impart medicinal property to organic woven/ knitted fabrics with combination medicinal herb extracts to care against common skin diseases. The highlight of this work is that all the stages of fabric processing were carried out using natural materials. The aqueous extract of the combinational medicinal herbs such as Neem, Turmeric, and Holy Basil were applied to the fabrics by means of the pad-dry-cure method using two pre-mordanting techniques with combination herbal extract concentrations as 33% owm and 100% owm each of constituent herb powder in the combination herbal extract formulation. Similarly, a combination of herbal extracts prepared from solvent extraction method was encapsulated onto the fabric as microcapsules with a combination herbal extract concentration of 10% owm. The treated fabric samples were tested for their antibacterial activity and comfort properties. The antibacterial and comfort property test results were statistically analysed using multivariable ANOVA with Design-Expert software. The antibacterial test results analysis shows that the treatment methods had a significant influence on the antibacterial activity; the type of fabric used had a marginally significant influence on the results. The treatment methods also had a significant influence on the antibacterial performance of the fabric after washing. In comfort property test results analysis, the treatment methods significantly influenced the water absorbency property; the type of fabric used significantly influenced the air-permeability, water absorbency, and drapeability properties. To confirm the health-care potential of the developed fabrics, nightwear apparels were developed from the best combination of herbal extract treated fabric based on its antibacterial and comfort properties. A wearer trial was conducted in a nature cure centre with the developed apparel and significant improvement noticed in all the cases.

Keywords : Medicinal herbs, combination herbal extract, Microencapsulation, Antimicrobial, Comfort, Health care.

14.1   Introduction

The World Health Organization (WHO) estimates that 4 billion people, 80 percent of the world population, presently use herbal medicine for some aspect of primary health care. Substances obtained from the plants remain the basis for a large proportion of the commercial medications used today for the treatment of various diseases. India's ancient form of medicine, *Ayurveda*, had a branch called Ayurvastra. Ayurvastra means the clothes made from organic cotton fabric that has been permeated with special herbs and oils that promote health and cure diseases.

In *Ayurveda*, single or multiple herbs (polyherbal) are used for the treatment. The *Ayurvedic* literature *Sarangdhar Samhita* emphasized the concept of polyherbalism to attain greater healing effectiveness. The active phytochemical components of individual plants are inadequate to achieve the needed therapeutic effects. When combining multiple herbs in a particular proportion, it will provide a better healing effect and decrease the toxicity [1].

The consumers are now conscious of the hygienic lifestyle and there is necessity and expectation for a wide range of textile products finished with antimicrobial properties [2]. In the present scenario of environmental consciousness, the new quality requirements not only emphasize on the intrinsic functionality, durability of the product but also the production process should be eco-friendly. Therefore, research on eco-friendly antimicrobial agents based on natural products for textile application is attaining worldwide attention. Different classes of active ingredients noticed in extracts of natural products [3]. The relatively lower incidence of negative effects of herbal products as compared to modern synthetic pharmaceuticals, coupled with their lower cost, can be exploited as an attractive eco-friendly alternative to synthetic antimicrobial agents for textile applications [4-5]. Natural dyes can be found with offering antifungal as well as antimicrobial finish to textiles as they are known for their dyeing and medicinal properties for a long time [6]. Health and hygiene are the primary requirements for human beings to live comfortably and work with maximum efficiency. To safeguard humankind from pathogens and to avoid cross infection, antimicrobial finish has become necessary. With the advent of new technologies, the rising needs of the consumer in the wake of health and hygiene can be satisfied without compromising the issues related to safety, human health, and the environment. Identifying new potential antimicrobial substances, such as chitosan from nature can considerably lower the undesirable activities of the antimicrobial products [7-9]. The skin sensorial comfort is negatively affected by hydrophobic, smooth (flat) surfaces that easily cling to sweat-wetted skin, or which tend to make textiles stiffer. As guidelines for the improvement of the thermo physiological or skin sensorial wear comfort, it is insisted to use hydrophilic treatments in a suitable concentration and spun yarns instead of filaments [10].

Hence, by considering the combination of the above aspects, herbal extracts treated apparel was developed for common skin diseases, Apart from this an attempt is made to provide it with more comfortable for the wearer by looking into the comfort aspects of the fabric in caring common skin diseases.

14.2   Materials & methods

14.2.1   Materials

14.2.1.1   Fabrics

The particulars of the organic woven/knitted fabrics used for the development of health care apparel are furnished in table 14.1.

Table 14.1: Fabric particulars used for the development of health care apparel

Particulars	Organic Woven fabric	Organic Knitted fabric
Fabric structure	Plain weave	Single Jersey
Yarn count (Ne)	40s Combed	40s Combed Hosiery
Ends/ Inch (or) Wales/inch	144	27
Picks/Inch (or) Course/inch	72	46
Grams/ Square metre	144.3	93.5

14.2.1.2   Medicinal Herbs & Natural materials

The particulars of the medicinal herbs and natural materials used for the development of health care apparel are furnished in table 14.2.

Table 14.2: Herbs & Natural materials used

Common name of the herbs used	Botanical name of the herbs used	Parts used
Turmeric	Curcuma longa	Rhizome
Neem	AzadirachtaIndica	Leaves, barks
Holy basil	Ocimum sanctum	Leaves
Banana tree	Musa Acuminata	Stem
Myrobalan	Terminaliachebula	Fruit
Soap nut	SapindusMukorossi	Shell

14.2.2   Methodology

14.2.2.1 Natural desizing

The woven fabric material is first wetted in a solution containing *Sapindusemerginata* seed extract (Soap-nut oil) 5% owm with an M: L ratio of 1:20 and left in this bath for 48 hours to carry out the natural desizing process.

14.2.2.2   Natural scouring

The Scouring process for woven and knitted fabrics was carried out in the solution containing 10% owm neem tree bark ash with a M:L ratio of 1:10, pH of 12.0 at 90° C for three hours and washed several times till the material is brought to neutral pH.

14.2.2.3   Natural bleaching

The scoured cotton fabrics (woven & knitted) are exposed to direct sunlight with the use of a natural grass base, which carries out the natural bleaching process through photolytic oxidation process.

14.2.2.4   Premordanting with myrobalan

Premordanting with dried myrobalan fruit powder extract 30% owm with M:L ratio as 1:20 at 60° C for 30 minutes and washed several times till the material is brought to neutral pH.

14.2.2.5   Premordanting with banana stem extract

Premordanting with banana stem extract 5% owm, which is a gum extracted by soaking one kg of banana stem in two litres of water for a day, with M:L ratio as 1:20 at room temperature for 60 minutes and washed several times till the material is brought to neutral pH.

14.2.2.6   Mordant free method

In this method, no mordants are used. Instead, this method is used to know the mordanting potential of the holy basil, which is already a constituent herb for the herbal combination obtained for this development work.

14.2.3   Herbal powder preparation

14.2.3.1   Neem, Holy basil leaf powder preparation

The fresh leaves of Neem and Holy basil were shadow dried and the dried leaves were then powdered using a pulverizer.

14.2.3.2   Turmeric powder preparation

The fresh rhizomes of Turmeric were shadow dried and the dried tubers were crushed manually into small pieces. The debris was removed and the remaining was powdered using a pulverizer.

14.2.4   Herbal extract preparation

14.2.4.1   Aqueous extraction method

The herbal extracts intended for herbal dyeing were prepared using aqueous extraction method, in which the herb powders are soaked in water over night in the ratio of 1:10 and boiled to get the extract.

14.2.4.2   Solvent extraction method

The herbal extracts intended for microencapsulation were prepared with methanol as a solvent to extract the active components from the herbs.

14.2.5   Application methods of combination herbal extracts

The combinationherbal extracts were applied to pre-treated woven and knitted fabrics using pad-dry-cure technique for the following processes.

- Herbal dyeing
- Micro-encapsulation

14.2.5.1   Herbal dyeing

The process parameters adopted for the herbal dyeing with pad-dry- cure method are given in table 14.3.

Table 14.3: Process parameters used in herbal dyeing

S. No	Description	Process parameters
1	Herbs used	Turmeric, neem and holy basil.
2	Combination Herbalextract concentration	i) 33% owm each of constituent herb powder ii)100%owm each of constituent herb powder
3	MLR	1:20
4	Temperature	80-90°C
5	Time	1 Hour
6	pH	8.5 – 9.0
7	Stabiliser	Neem ash(5% owm)
8.	Drying & Curing	Room temperature

After dyeing, the fabric samples were washed several times until the material attains neutral pH. To enhance the durable fixation of the dye substance over the fabric, an after treatment was given with banana stem extract (3% owm) which acts as a gum to fix the herbal substance on the fabric with an M: L ratio of 1:5 for 5 minutes duration at room temperature. After dyeing, the material was washed several times to attain neutral pH. After washing, the samples were dried & cured in shadow at room temperature to retain the medicinal properties.

14.2.5.2   Microencapsulation

Microcapsules were prepared to employ sodium alginate. 100 ml of 3% sodium alginate was prepared. Then 1.5 gram of methanol extract of combination herbs (Neem, Turmeric & Holy basil) mixed in equal proportion was added to the polymer solution and mixed thoroughly to form smooth viscous dispersion. This was sprayed into a calcium chloride solution by means of a sprayer. The droplets were retained in calcium chloride for 15 minutes. The microcapsules were obtained by decantation and repeated washing with iso propyl alcohol followed by drying at 45°C for 12 hours. The microcapsules were then used for finishing on the selected fabrics. The observation of microcapsules is shown in Fig 14.1. The process parameters adopted for the microencapsulation process is given in Table 14.4.

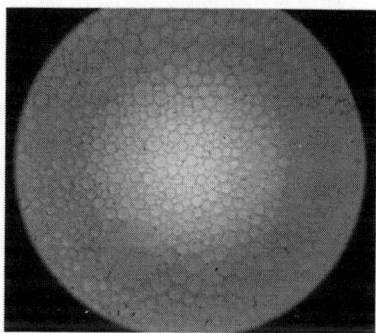

Figure 14.1: Observation of Microcapsules under Microscope

Table 13.4: Process parameters used in Micro-encapsulation

S. No	Description	Process parameters
1	Herbs used	Turmeric, neem and holy basil.
2	Combination Herbalextract concentration	10% owm
3	Chemicals used	Sodium Alginate - 3% Calcium Chloride - 2%
4	Binder	Citric acid - 8%
5	MLR	1:20
6	Drying& Curing	Room temperature

14.2.6   Experiment design

The experiment design particulars for the combination herbal extract treatment are shown in Table 14.5.

Table 14.5: Experiment design particulars

S. No	Description	Parameters
1.	Software used	Design Expert
2.	Experiment Design	General Factorial Design
3.	No.of Factors	Type of material, Treatment method,Combination Herbal extract concentration
4.	No. of levels	
	Type of material	Woven (W) & Knitted (K)
	Treatment methods	Myrobalan Pre-mordanting (MM), Banana Stem Pre-mordanting (BM), Microencapsulation (ME) & Mordant free (NM)
	Combination herbal extract concentration	33% owm & 100% owm each of constituent herb powder (Aqueous Extract); 10% owm for microencapsulation (ME)
5.	No.of runs	14

2.7   Testing of healthcare fabrics

2.7.1   Antibacterial tests

The Antibacterial test was carried out using AATCC 147& AATCC 100 Test Methods. The wash durability test was carried out using AATCC test method 61-2009. Subsequently, the samples were tested for their bacterial reduction activity once in every 5 washes from 0-30 washes for all the samples.

2.7.2   Fabric surface characterisation tests

The confirmation of the presence of active components was carried out using Fourier Transform Infrared Spectroscopy (FTIR) Tester.The surface analysis of the treated fabrics was carried out using SEM (Scanning Electron Microscopy) Tester.

14.2.7.3   Comfort property tests

The air permeability test of the fabric was carried out in KES Air-permeability tester (KES-F8-AP1) using the ASTM D737-96 Standard Test Method. The water absorbency test was carried out using water absorbency test (AATCC/ ASTM TEST METHOD TS-018). The thermal conductivity test of the fabric samples was conducted using lee's disc method. ASTM method E 96-80 test procedure was used for the measurement of water vapour permeability. The drapeability of the developed samples was tested with drape meter by using Cusack drape test method.

14.2.8   Development of health care apparel for skin diseases

After the testing of developed samples for antimicrobial and comfort properties, the sample which excelled in both aspects with cost-effectiveness was selected and a nightwear was developed.

14.2.9   Wearer trial

The performance of the developed health care apparel was evaluated by conducting a wearer trial in a nature cure center by giving it to patients with various skin diseases.

14.3   Results & discussions

14.3.1   Assessment of antibacterial activity of fabrics – AATCC 147 – Agar

Diffusion method (Qualitative Test)

The antibacterial activity test results of the developed samples using the agar diffusion test method (**AATCC-147)** are given in the table 14.6.

Table 14.6: Assessment of antibacterial activity by agar diffusion method (Qualitative test)

S. No	Sample Description	Antibacterial activity (Zone of Bacteriostasis – mm)	
		Escherichia Coli	*Staphylococcus aureus*
1	MM 100% owm (W)	29	35
2	MM 100% owm (K)	0	37
3	BM 100% owm (W)	0	34
4	BM 100% owm (K)	0	29
5	NM 100% owm (W)	0	0
6	NM 100% owm (K)	0	30
7	MM 33% owm (W)	0	32
8	MM 33% owm (K)	0	44
9	BM 33% owm (W)	0	25
10	BM 33% owm (K)	0	40
11	NM 33% owm (W)	26	30
12	NM 33% owm (K)	35	39
13	ME 10% owm (W)	0	0
14	ME 10% owm (K)	0	0

The test result shows the observations for the zone of Bacteriosis after 24 hours of incubation. The result shows that Bacterial activity against *Staphylococcus aureus* far exceeds the activity against *Escherichia coli*. It assures that all the samples have the ability to cure the skin diseases caused by *S.aures* bacteria. *S.aures* bacteria are frequently part of the skin flora found in the nose and on skin. About 20% of the human populations are long-term carriers of *S. aureus*. *S. aureus* can cause a range of illnesses from minor skin infections, such as pimples, impetigo, boils (furuncles), cellulitis folliculitis, carbuncles, scalded skin syndrome, and abscesses.

Antibacterial activity was observed against both *E.coli* and *S.aureus* for samples produced without mordant (Woven and knitted) with an herbal concentration of 33% owm of constituent herb powder and myrobalan pre-mordanted (knitted) with an herbal concentration of 100% owm of constituent herb powder in the combination herbal extract. These samples have the ability to fight against the diseases caused by not only the S.aureus but also fight against the diseases caused by *E. coli* such as pneumonia, meningitis, infected bones and joints, and skin and soft tissue infections (especially in people who have diabetes).

The multivariable ANOVA results show that the treatment method had a significant influence on antibacterial activity. However, the type of fabric and combination herbal concentration used had no significant influence on the antibacterial activity. And also no significant influence observed with respect to the interaction of factors.

14.3.2   Assessment of antibacterial activity by bacterial reduction test (AATCC-100)

Table 14.7: Assessment of antibacterial activity by bacterial reduction test (AATCC-100)

S. No	Sample	Bacterial reduction (%)	
		Escherichia Coli	*Staphylococcus aureus*
1.	MM 33% owm (K)	93	91
2.	MM 100% owm (K)	52	100
3.	NM 33% owm (W)	>99.99	>99.99
4.	NM 33%owm (K)	>99.99	>99.99
5.	ME 10% owm (K)	>99.99	>99.99

To confirm the qualitative analysis test results obtained from AATCC 147, a confirmative antimicrobial test was carried out using the challenge test method (AATCC-100) for the selected samples and the results are given in Table 14.7.

The test results clearly confirm that all the selected samples have very good antibacterial activity against both gram-positive and gram-negative bacteria.

Table 14.8 shows the wash durability test results of the treated samples. The result shows a declining trend in percentage reduction in bacteria was observed with washing duration. More antibacterial activity was recorded in microencapsulated samples as compared to the others at the end of the 30th wash. The above results indicate that the herbal molecule has a good diffusion rate into the yarn structure and its effectiveness in controlling bacterial activity for prolonged periods. The multivariable ANOVA results show that the treatment methods had a significant influence on the wash durability. And also no signuficant influence observed with respect to the interaction of factors.

The pre-mordanted and micro-encapsulated samples have shown good wash durability when compared to the mordant free method. In case of pre-mordanting process with myrobalan and banana stem extract which is having proven mordanting ability, they were treated with 3% extract of banana stem extract after dyeing, which helps the herbal molecules to firmly fix onto the fabric. This might be perhaps the reason for most wash durability (up to 25 washes). In the case of micro-encapsulated samples, the use of citric acid binder helps to achieve the durability against washing by firmly fixing the microcapsules on the fabric.

Table 14.8: Wash durability test results

Washes	Bacterial reduction (%)							
	Myrobalan Pre-mordanted		Banana stem Pre-mordanted		Mordant free Method		Micro Encapsulation method	
	E.coli	S. aureus	E.coli	S. aureus	E.coli	S. aureus	E.coli	S. aureus
Raw Sample	99.99	63.84	99.99	69.32	99.99	99.99	99.99	99.99
after 5 washes	77.91	63.15	79.01	65.62	75	68.9	83	80.21
after 10 washes	62.79	60.14	0	61.37	14	8.53	74.67	69.68
after 15 washes	28.49	59.59	0	35.62	0	0	63.63	60.05
after 20 washes	18.02	40.82	0	14.25	0	0	51.58	49.69
after 25 washes	7.82	23.87	0	5.56	0	0	42.36	39.26
after 30 washes	0	0	0	0	0	0	31.89	28.99

14.3.4   Analysis of FTIR (Fourier Transform Infrared Spectroscopy results)

The Fourier Infrared Spectroscopy (FTIR) images of parent knitted fabric and knitted microencapsulated samples are given in figure 14.2 & 14.3 respectively. In parent knitted fabric, Cellulosic–OH Stretching is observed at around at 3360 cm $^{-1}$. Aliphatic C-H stretching is observed at 2901- 2914 cm^{-1}. In the microencapsulated knitted fabric treated with 10% owm of combination herbal extract, in addition to the peaks appeared in parent fabric, a new peak corresponding to C=O(Carbonyl) stretching is observed at1724 cm^{-1} and this may be due to the incorporation of active components of the combination herbal extract. Other peaks also observed at 707.90 cm $^{-1}$ and 1236.41 cm $^{-1}$ and stretching occurred at 2360.95 cm-1 & 2343.59 cm-1, which may be due to the incorporation of the active components of the combination herbal extract used in the fabric.

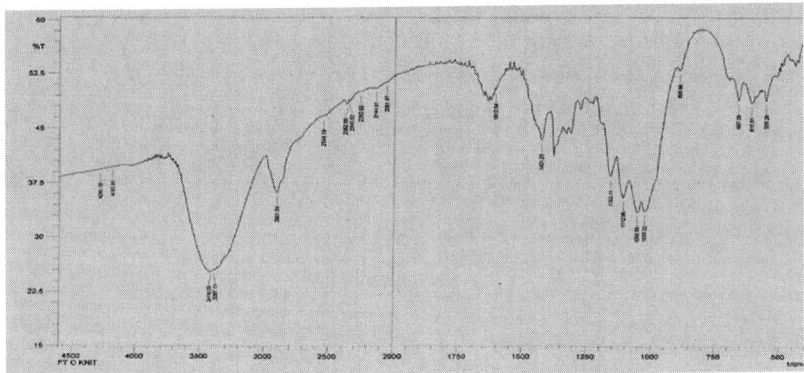

Figure 14.2: Parent knitted fabric

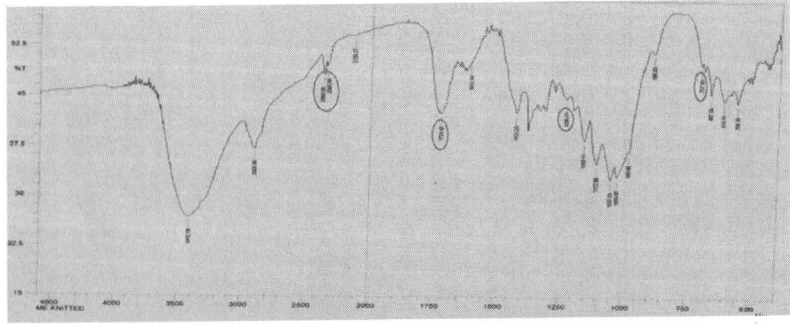

Figure 14.3: Microencapsulated knitted fabric

14.3.5   Analysis of SEM (Scanning Electro Microscopy) results

The scanning electron microscopy (SEM) images of parent knitted fabric and treated fabric samples are shown in figure 14.4 to 14.7. From the SEM photographs, it is clearly seen that the presence of the applied herbal substances firmly attached to the fabric in both micro-encapsulated fabric sample and the sample pre-mordanted with myrobalan.

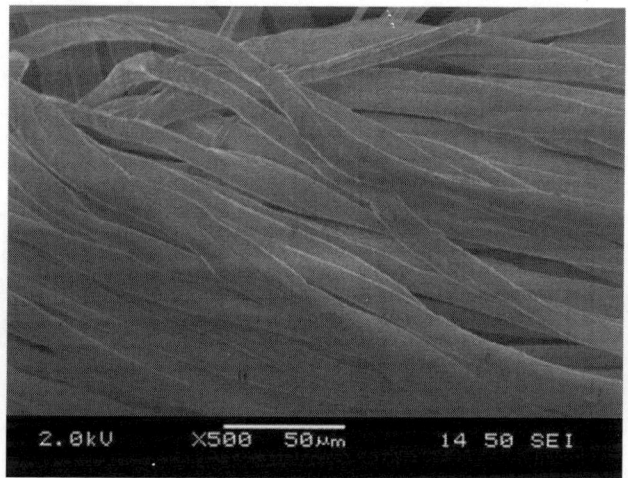

Figure 14.4: SEM image of parent knitted fabric sample

Figure 14.5: SEM image of Microencapsulated knitted fabric

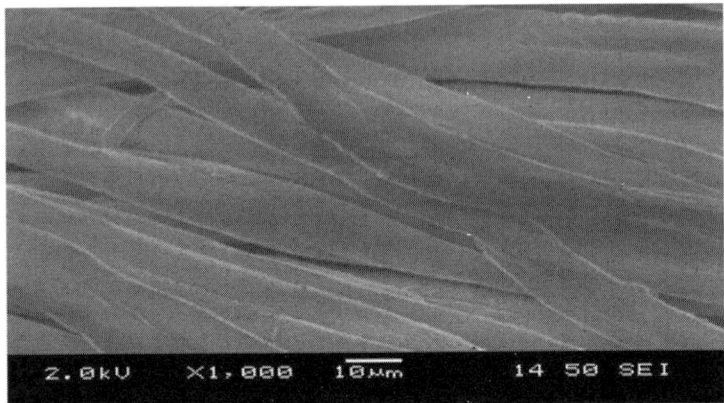

Figure 14.6: SEM image of parent knitted fabric sample

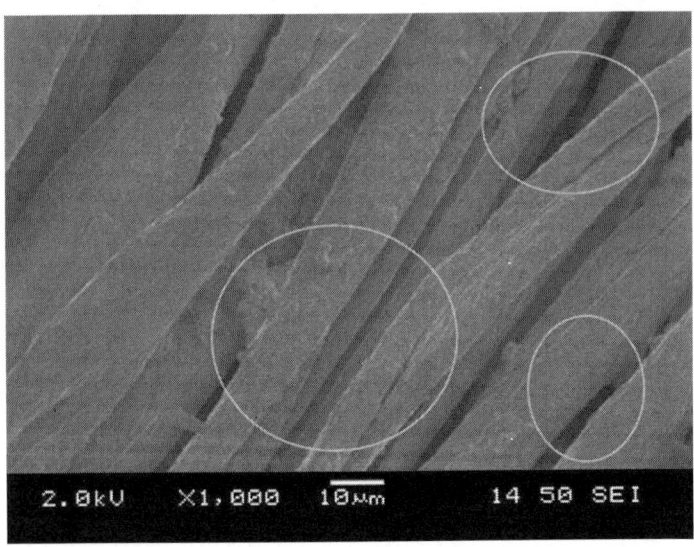

Figure 14.7: SEM image of Myrobalanpre-mordantedfabric sample treated with 33% owm of constituent herb powder in the combination herbal extract

14.3.6.   Effect of combination of herbal extract treatment on comfort properties

The fabric samples were tested for five main aspects in relation to wearer's comfort. They are air permeability, water absorbency, drape, water vapor permeability and thermal conductivity.

14.3.6.1 Evaluation of air-permeability characteristics of the treated samples.

The air permeability results of the samples are presented in Table 14.9

Table 14.9: Analysis of air permeability results

S.No	Sample	Air Resistance (Kpa s/m)	Air Permeability (cc/s/sq.cm)
1	MM 100% owm (W)	1.7	7.33
2	MM 100% owm (K)	0.14	88.96
3	NM 100% owm(W)	1.47	8.47
4	NM 100% owm(K)	0.15	83.03
5	BM 100% owm(W)	1.9	6.56
6	BM 100% owm(K)	0.14	88.96
7	MM 33% owm(W)	1.58	7.88
8	MM 33% owm(K)	0.14	88.96
9	NM 33% owm(W)	1.72	7.24
10	NM 33% owm(K)	0.15	83.03
11	BM 33% owm(W)	2.02	6.17
12	BM 33% owm(K)	0.36	34.59
13	ME 10% owm(W)	3.37	3.70
14	ME 10% owm (K)	0.16	77.84

From the test results, it is obvious that the knitted fabric samples, which are with single jersey knit structure offered a higher air-permeability level than woven fabrics. This property can be exploited by developing the apparel with the knitted samples for more complex skin diseases such as allergic dermatitis, Scabies, Eczema, Psoriasis, etc. since in these diseases, the fabric surface friction with skin may aggravate the diseases. This property also helps to dry out the sweat emitting from the skin easily, which will give wearing comfort to the wearer. The multi-variable ANOVA results reveal that the type of fabric used has a significant influence on the results. But neither the application method nor the combination herbal extract concentration used had no influence on the results and no significant influence observed with respect to the interaction of factors.

14.3.6.2   Evaluation of water absorbency characteristics of the treated samples

The water absorbency of the fabric samples is presented in table 14.10. Greater the time taken, lesser is the water absorbency. In that respect, most knitted samples have a higher rate of absorption than the woven samples, and the microencapsulated samples had the least absorbency which may be due to the hindrance of the microcapsules in allowing the water to contact with the fabric hydrophilic surface. The lower water absorbency time of the most knitted fabric samples helps to absorb the sweat/ secretions from the skin at a faster rate which will lead to wearing comfort to the wearer.

Table 14.10: Water absorbency test results

S. No	Sample	Average Absorbency time (seconds)
1	NM 33%o wm (W)	3.41
2	NM 33% owm (K)	2.37
3	MM 33%owm (W)	5.05
4	MM 33 % owm (K)	3.14
5	BM 33% owm (W)	4.14
6	BM 33% owm (K)	2.89
7	ME 10% owm (W)	8.22
8	ME10% owm (K)	7.55
9	NM 100% owm (W)	3.52
10	NM 100% owm (K)	2.51
11	MM 100% owm (W)	3.35
12	MM 100% owm (K)	2.31
13	BM 100% owm (W)	4.83
14	BM 100%owm(K)	3.34

The multivariable analysis results reveal that the type of fabric used and treatment methods had a significant influence on the results. But the combination herbal extract concentration used had no influence on the results and no significant influence observed with respect to the interaction of factors.

14.3.6.3   Evaluation of drape coefficient of the treated samples

The drape co-efficient of the treated samples is given in Table 14.11

Table 14.11: Drapeability test results

S. No	Sample	Drape co-efficient
1	MM 100%owm (W)	0.69
2	MM 100% owm (K)	0.34
3	NM 100% owm (W)	0.74
4	NM 100% owm (K)	0.46
5	BM 100% owm (W)	0.77
6	BM 100% owm (K)	0.38
7	MM 33% owm (W)	0.36
8	MM 33% owm (K)	0.35
9	NM 33% owm (W)	0.68
10	NM 33% owm (K)	0.38
11	BM 33% owm (W)	0.65
12	BM 33% owm (K)	0.50
13	ME 10% owm (W)	0.63
14	ME 10% owm (K)	0.35

From the test results, it is evident that knitted fabrics exhibited higher drapability than woven fabrics. This property can be positively exploited in the development of health care apparel since the fabric responds easily to the wearer body contour due to its good drapeability nature. The study reveals that the type of fabric used had a significant influence on the results. But both treatment methods and the combination herbal extract concentration used had no significant influence on the results and also no significant influence observed with respect to the interaction of factors.

14.3.6.4   Evaluation of water vapour permeability characteristics

The water vapor permeability test results of the treated fabrics are given in

Table 14.12

Table 14.12: Water vapor permeability test results

Treatment method	Water Vapour Permeability (Grams/sq.metre/Day)					
	Woven 33% owm	Knitted 33% owm	Woven 100% owm	Knitted 100% owm	Woven 10% owm	Knitted 10% wm
Myrobalan Pre-mordanted (MM)	1437.54	1603.42	1842.40	1491.46	–	–
Banana stem Pre -mordanted	1769.29	1603.42	1437.544	1658.71	–	–
Mordant free	1491.46	1769.29	1382.25	1492.83	–	–
Micro-encapsulated	–	–	–	–	1271.67	1603.42

A fabric of low moisture vapour permeability is unable to pass sufficient perspiration and this leads to sweat accumulation in the clothing and hence discomfort. The test results show that all the treated samples are possessed with very good water vapour permeability characteristics. This property is an essential property for the health care apparel since the evaporation of the sweat from the skin should be faster to attain wearing comfort and to have a very good thermo-regulation function of the skin. If the water vapour permeability is poor, then it will affect the thermoregulation function of the skin. There are two main properties of clothing, which affect the handling of moisture. Firstly, there is the ease with which clothing allows the perspiration to be evaporated from the skin surface during the activity. Secondly, after the activity has ceased, there is a need for the moisture that is contained in the clothing layer next to the skin to dry out quickly. This ensures that the wearer does not lose heat unnecessarily through having wet skin [11].

The multivariable analysis results reveal that there is no significant influence with respect to the type of fabric used, treatment methods and combination herbal extract concentration and no significant influence observed with respect to the interaction of factors.

14.3.6.5   Evaluation of thermal conductivity characteristics of the treated Samples

The thermal conductivity test results of the treated samples are given in Table 14.13.

Table 14.13: Thermal conductivity test results

S. No	Sample	Thermal conductivity (w $k^{-1}m^{-1}$)	
		Woven	**Knitted**
1	Control fabric	5.7×10^{-4}	6.87×10^{-4}
2	NM 33% owm	5.244×10^{-3}	5.49×10^{-3}
3	MM 33% owm	8.06×10^{-4}	6.13×10^{-3}
4	BM 33% owm	7.11×10^{-4}	9.42×10^{-3}
5	NM 100% owm	5.83×10^{-4}	6.20×10^{-3}
6	MM 100% owm	6.21×10^{-4}	7.69×10^{-3}
7	BM 100% owm	5.94×10^{-4}	6.48×10^{-3}
8	ME 10% owm	6.11×10^{-4}	6.12×10^{-3}

From the test results, it is evident that the knitted fabric samples have a higher thermal conductivity than woven materials. If a person is comfortable (that is, in heat balance) at rest then a burst of hard exercise will mean that there is a large amount of excess heat and also perspiration to be dissipated. Clothing has a large part to play in the maintenance of heat balance as it modifies the heat loss from the skin surface and at the same time has the secondary effect of altering the moisture loss from the skin. However, no one clothing system is suitable for all occasions: a clothing system that is suitable for one climate is usually completely unsuitable for another. In our case, knitted fabric samples, which are having a marginally significant variation in thermal conductivity characteristics when compared to woven fabrics can be used for summer occasions where the heat exchange process is more and woven fabrics can be used in winter seasons where preservation of body heat is essential to maintain the thermo-regulation function. The knitted samples treated with banana stem pre-mordanting technique with a combination herbal extract concentration of 33% owm of constituent herb powder in the combination herbal extract formulation exhibited a higher thermal conductivity.

The multivariable analysis results reveals that the type of fabric used has a marginally significant influence on the results. But neither the application method nor the combination herbal extract concentration used had no influence on the results and also no significant influence observed with respect to the interaction of factors.

14.3.7 Evaluationof curative performance of ecofriendly health care apparel using wearer trial

The wearer trial was conducted in a nature cure hospital. The knitted fabric sample pre-mordanted with myrobalan and treated with 33% owm of constituent herbs in the combination herbal extract formulation was selected based on its antibacterial property, comfort properties, and nightwear apparels were developed. The developed was tested for its curative performance and comfort properties by giving it to patients with various skin diseases. Based on the doctor's evaluation, a significant improvement was noticed in all the cases and no side effects or allergies were experienced by the patients.

14.3.7.1   *Wearer Trial summary*

The results of the wearer trial for the nightwear apparel are given in Table 14.14. The wearer trial results shows that the product has not caused any adverse effects (Like rash or itch). The idea of medically treated apparel impressed the patients greatly. The lists of patients with a history of skin disease were given with the apparel and the doctor assessed its effects periodically.

The feedback from the patients shows that the apparel has a good acceptance. The bio-chemical analysis of the herbs used for the development of the health care apparel can be elucidated and its exact role in the curing of the diseases can be well established.

Table 14.14: Wearer Trial summary

Patient	Age	Disease	Curative performance
1	25	Psoriasis	Significant
2	27	Dermatitis	Significant
3	42	Eczema	Significant
4	38	Sunburn	Significant

14.4   Conclusions

The ayurvedic concept of the polyherbal formulation was adopted in this work to develop combination herbal extract treated healthcare apparel for common skin diseases using the medicinal herbs neem, turmeric, and holy basil.

The antibacterial test results analysis of the combination herbal extracts treated fabrics shows that only the treatment methods had a significant influence on the antibacterial activity, the fabric used had a marginally significant influence on the results and the herbal concentration had no significant influence on the results. The wash durability-bacterial reduction results analysis shows that the treatment methods had a significant influence on the wash durability results with micro-encapsulated samples lasted with activity even after 30 washes and significant drop noticed in bacterial reduction% for all the samples between wash cycles from 0-30.

The comfort property test results analysis of the combination herbal extracts treated fabrics shows that in air-permeability, drapeability tests, and all the knitted fabric samples have shown good performance and neither treatment methods nor herbal concentration had no significant influence on the above-mentioned properties in all the samples. In a thermal conductivity test, marginally significant variation noticed with respect to the type of fabric used and knitted fabric samples has good thermal conductivity than woven fabric samples. In water absorbency test, except microencapsulated samples, all the other samples have shown good performance and a significant variation with respect to fabric used and treatment methods was noticed. In water vapour permeability test, no significant variation observed with respect to the type of fabric used, treatment method and herbal concentration.

In FTIR (Fourier Transform Infrared Spectroscopy) analysis, the presence of active components had been confirmed. The SEM (Scanning Electron Microscopy) analysis also confirmed the presence of the applied herbal substances on the fabric structure.

The knitted fabric sample pre-mordanted with myrobalan and treated with 33% owm of constituent herbs in the combination herbal extract formulation was selected based on its anti-microbial property, comfort properties, and nightwear apparels were developed.

To confirm the healthcare potential of the developed apparel, a wearer trial was conducted in a nature cure centre by giving the developed nightwear apparel to the patients suffering from common skin diseases. Significant improvement was observed in the curative performance of the developed apparel in all the cases through Doctor's evaluation.

14.5   References

1. Parasuraman, S, Thing, GS & Dhanaraj, SA 2014, 'Polyherbal formulation: Concept of Ayurveda', Pharmacognosy reviews, Vol. 8(16), pp. 73-80.

2. Ramachandran T, Rajendrakumar K &Rajendran R, 2004, "Antimicrobial finishes – An overview", IE *Journal*, Vol.84, pp 42-47.

3. ThilagavathiG & Kannaian T, 2008, "Application of Pricky chaff (*Achyranthesaspera Linn.*) leaves as herbal antimicrobial finish for cotton fabric used in health care textiles", *Natural product radiance*, Vol. 7(4), pp 330-334.

4. Joshi M, Wazed Ali S &Purwar R, 2009, "Ecofriendly antimicrobial finishing of textiles using bioactive agents based on natural products", Indian *Journal of Fibre& Textile Research*, Vol.34, pp 295-304.

5. Ashis Kumar Samanta & Priti Agarwal, 2009, "Application of natural dyes on textiles",*Indian Journal of Fibre& Textile Research*, Vol. 34, pp 384-399.

6. Dhara Bajpai and Padma S Vankar, 2007, "Antifungal Textile Dyeing with *Mahonianapaulensis D.C.* Leaves Extract Based on Its Antifungal Activity", *Fibres and polymers*, Vol.8, No.5, pp 487-494.

7. Thilagavathi G & Krishna Bala S,2007, "Microencapsulation of herbal extracts for microbial resistance in healthcare textile", Indian *Journal of Fibre& Textile Research*, Vol. 32, pp 351-354.

8. Shanmugasundaram OL, 2007, "Antimicrobial finish in textiles", *The Indian Textile Journal*,.Vol.117 (11), pp 53-58.

9. Deepti Gupta, 2007, "Antimicrobial treatments for textiles", *Indian Journal of Fibre& Textile Research*, Vol. 32 , pp 254-263.

10. U.C. Hipler, P.C.Elsner, 2006, "Current problems in dermatology: Bio functional textiles and skin", S. Karger AG, Switzerland.

11. B.P. Saville, "Physical testing of textiles", 1999, Wood head Publishing Limited, UK.